THE PORT OF ROTTERDAM

MAERSK LINE

ect
ect
ect
MORTEN MÆRSK
DRAGØR

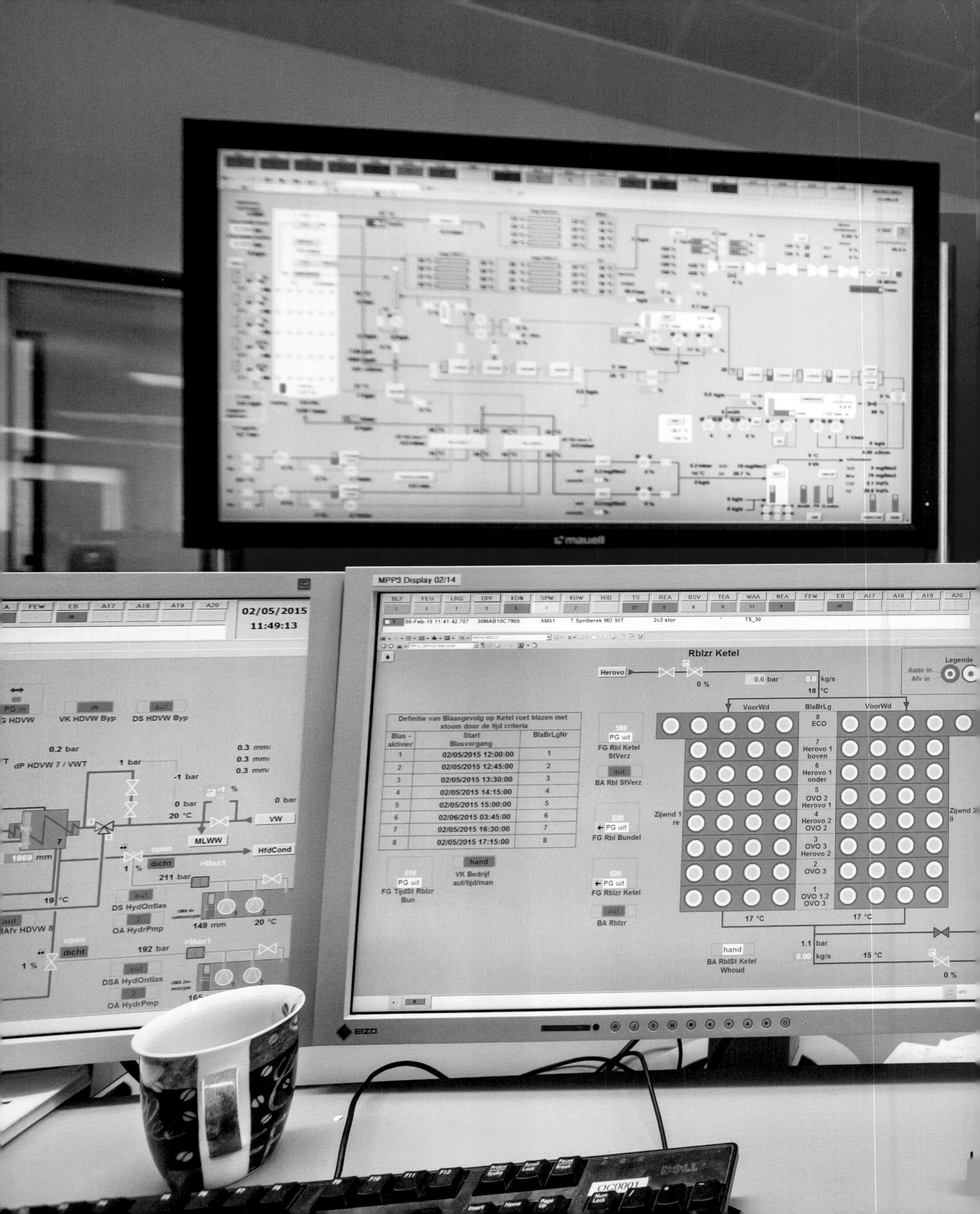

02/05/2015
11:49:13
MPP3 Display 02/14
Rblzr Ketel
Herovo
0.6 bar
0.0 kg/s
18 °C
0 %
Legende
VoorWd
BlaBrLg
VoorWd
Definitie van Blaasgevolg op Ketel roet blazen met stoom door de tijd criteria
Blas - aktivier
Start Blasvorgang
BlaBrLgNr
1 02/05/2015 12:00:00 1
2 02/05/2015 12:45:00 2
3 02/05/2015 13:30:00 3
4 02/05/2015 14:15:00 4
5 02/05/2015 15:00:00 5
6 02/06/2015 03:45:00 6
7 02/05/2015 16:30:00 7
8 02/05/2015 17:15:00 8
PG uit
FG Rbl Ketel StVerz
BA Rbl StVerz
PG uit
FG Rbl Bundel
PG uit
FG Rblzr Ketel
BA Rblzr
PG uit
FG TijdSt Rblzr Bun
hand
VK Bedrijf aut/tijd/man
hand
BA RblSt Ketel Whoud
8 ECO
7 Herovo 1 boven
6 Herovo 1 onder
5 OVO 2 Herovo 1
4 Herovo 2 OVO 2
3 OVO 3 Herovo 2
2 OVO 3
1 OVO 1.2 OVO 3
Zijwnd 1 re
17 °C
17 °C
1.1 bar
0.00 kg/s
15 °C
0 %
VK HDVW Byp
DS HDVW Byp
dP HDVW 7 / VWT
0.2 bar
1 bar
-1 bar
0 bar
20 °C
VW
MLWW
HfdCond
dicht
211 bar
19 °C
DS HydOntlas
OA HydrPmp
149 mm
20 °C
dicht
192 bar
DSA HydOntlas
OA HydrPmp
EIZO

MPP3 Display 03/14
02/05/2015
11:49:14
Watercentrum
StoomSys
Eenheidleiding
FG algemeen installatie
FG Q MonstNa OndStel 1
FG Q MonstNa OndStel 2
FG Q MonstNa OndStel 3
Temporary connected for commissioning
EIZO
ABB
Corporate Express
Zoetjes
OC00
User Name

Marinke Steenhuis ed.

nai010 publishers

World Between City and Sea

THE PORT OF ROTTERDAM

text:
Marinke Steenhuis
Peter de Langen
Frank de Kruif
Lara Voerman
Isabelle Vries
Peter Paul Witsen

photography:
Jannes Linders
Siebe Swart

infographics:
Beukers Scholma

M
MSTX7

kpn
Deloitte

CONTENTS

OVERVIEW

CONTENTS

PORT PLACES

source: Port of Rotterdam Authority 2015

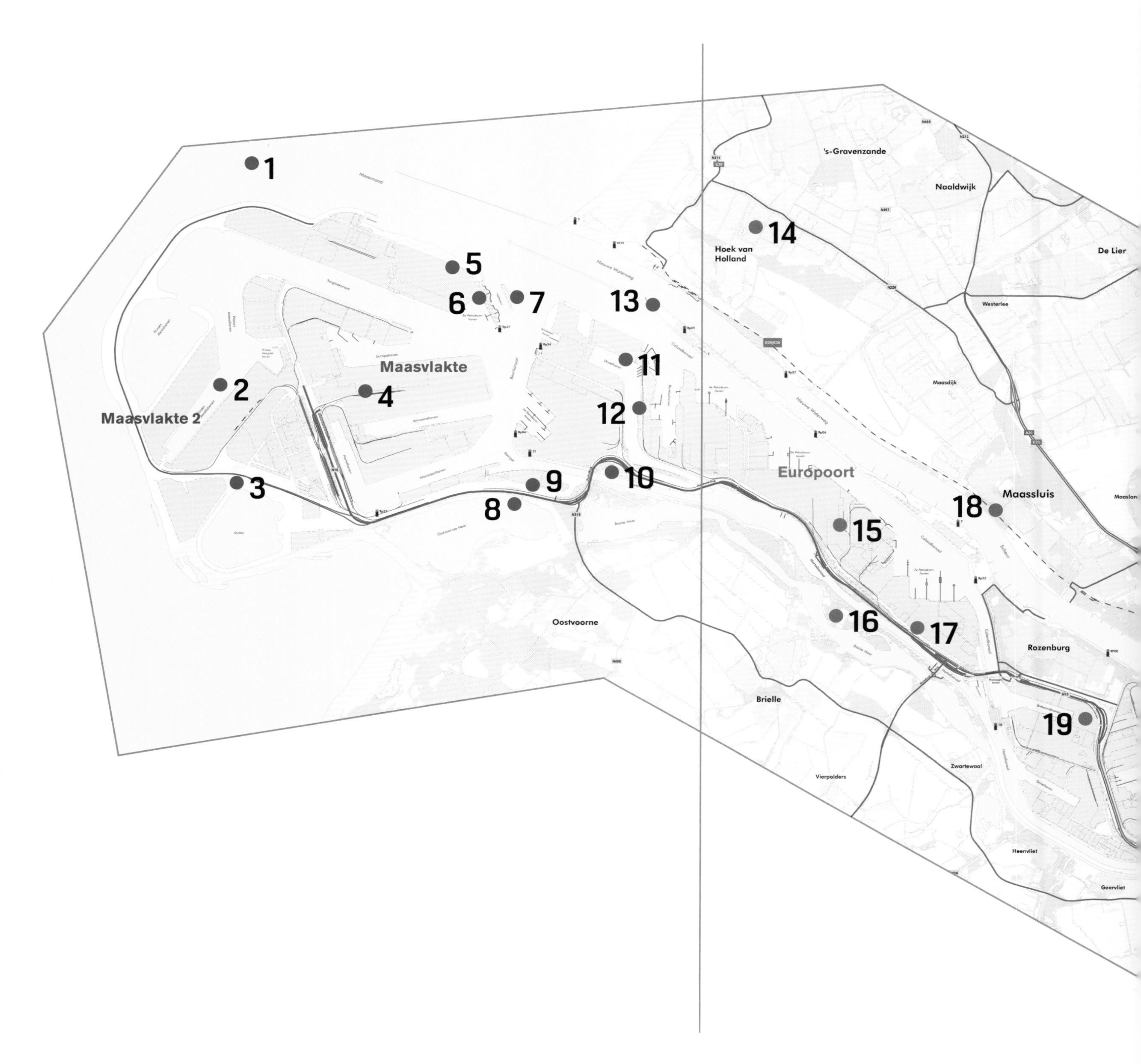

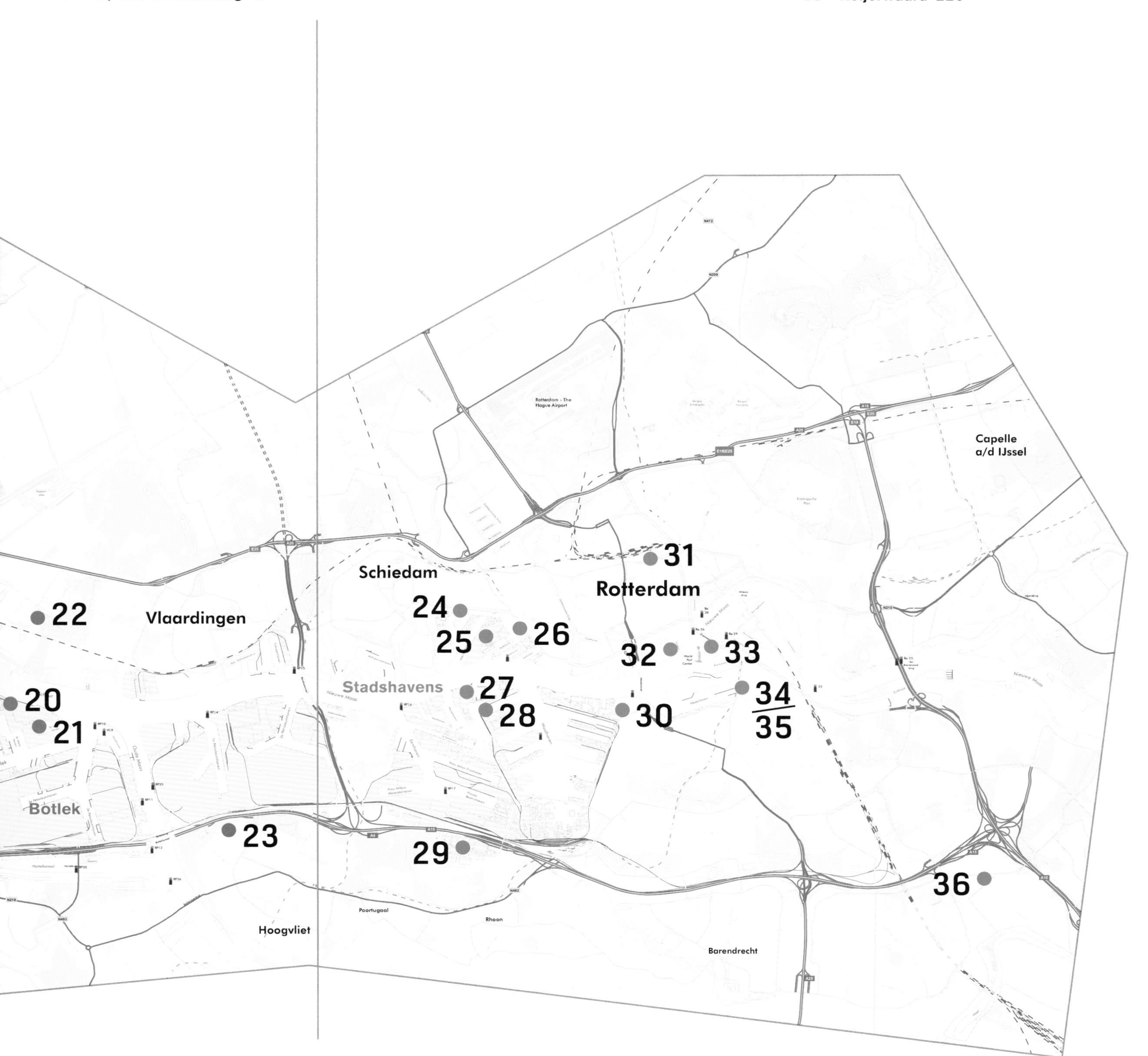

Capelle a/d IJssel
31
Schiedam
Rotterdam
24
22
Vlaardingen
25
26
32
33
Stadshavens
27
34
35
20
28
30
21
Botlek
23
29
36
Hoogvliet
Poortugaal
Rhoon
Barendrecht

FOREWORD

This book is as Rotterdam as it can be. It starts with the title, which indicates exactly what the book is about. And in the 270 pages that follow, this book does what the title promises, namely by articulating and depicting from different perspectives what happens in the port, how the port has developed, and how the port and the city continue to influence each other. The City of Rotterdam derives its identity from the port: it is straightforward, deeds rather than words, rolled-up sleeves, international, making money. All the clichés are true. And just as the identity of the city is determined by the port, the port cannot do without the city. Together they form a dual entity. Precisely at a time when more and more port work is coming in the form of (business) services, and when innovation is essential, it is important for the city to be attractive, creative and dynamic.

What is immediately striking about this book is the large number of aerial photos. They make the scale of the area clear. From the Van Brienenoord Bridge to the Maasvlakte 2 is 50 km: 10 km of city, and 40 km of port. From the centre of Rotterdam to the Maasvlakte 2 is as far as from Leiden or Breda. You have to see it from the air to be able to understand this clearly.

But the aerial photographs do not allow you to actually experience the port. They will not let you know what it feels like to see, from the 'Balcony of Europe' (Slag Maasmond), ships gliding past just 400 m from your nose under wind force 6 conditions, or to see from the end of the Oude Maasweg how tugboats manoeuvre a drilling platform into the Botlek, or to take the ferry from the Hook of Holland to the Maasvlakte, to sail along the large container terminals on the way to the FutureLand information centre, to take the Aqualiner from the Erasmus Bridge to Heijplaat, and to see how the RDM area has begun its second life. The port is changing constantly, and there is always something to experience. Let this wonderful book be an inspiration to discover that for yourself.

Allard Castelein
CEO Port of Rotterdam Authority

In 2009, when I had just become the Mayor of Rotterdam, I flew over the Rotterdam-The Hague metropolitan area for the first time. From the helicopter, I saw the urban region, interspersed with green villages. The water, rails, and bike paths looked like a dense network that had been projected atop the area. I saw businesses and offices in all shapes and sizes, from petrochemical facilities to greenhouses and government buildings. This area, which contains 2.2 million inhabitants, the port, the Westland and the city of diplomacy and peace, forms the economic heart of the southern Randstad region. In this metropolitan area, 15 per cent of the total Dutch income is earned.

So much for objective observations. But what I also noticed, and what even surprised me, was the beauty of the landscape. It kind of reawakened the poet in me; in the distant past, I had dreamed of becoming a poet. I saw the river, the lifeline of Rotterdam, meandering through the landscape like a connecting theme, sprinkled with multicoloured cranes and containers along the docks, and with the storage tanks acting like white dots on the i. Along the water was an endless procession of ships of all shapes and sizes, slicing through the Maas in their own poetic metre. The port of Rotterdam: an ingeniously constructed landscape with water, greenery, and swaying trees.

I am not sure, but I suspect that the port of Rotterdam is one of the few global ports where you can take beautiful bike rides and go on wonderful hikes, and where people on folding chairs sit along the banks of the river, enjoying the ships. And that might well be the secret that makes the Mainport of Europe such an excellent port: this combination of business activity and humanity. The port has the beauty of both a mathematical formula and a picturesque landscape painted by Johan Hendrik van Mastenbroek. Browse through this book, and you will understand what I mean. Excellence and beauty are two sides of the same coin.

Ahmed Aboutaleb
Mayor of Rotterdam

APL

Marinke Steenhuis

THE PORT AS A LANDSCAPE

The Port of Rotterdam is a world between the city and the sea, an industrial port landscape stretching for 40 km with a language and mentality all its own. The shipping routes that begin at the Port of Rotterdam extend around the world. This port is an economic network spanning the globe, stretching from the Russian oilfields to the Suez Canal, to China, to Brazil and the Panama Canal. The Port of Rotterdam also extends through the hinterland to the subsidiary ports of Coevorden, Delfzijl, Moerdijk and Alphen aan den Rijn, where transhipment centres for rail and shipping cargo ensure that the Rotterdam port machinery is kept well oiled. Oil is pumped through underground pipes as far afield as Flushing, Antwerp and the Ruhr Region. The heat generated from waste incineration is captured and channelled to inner-city Rotterdam. In the Westland district north of the port, the carbon dioxide released by port industries is used to grow greenhouse crops.

This book is about Rotterdam, and about how the port and the city have influenced each other; about the innovations in transport and transhipment that led again and again to a new type of port landscape. The revenues generated by steam shipping made it a sensible proposal in 1872 to invest in the New Waterway, and the invention of oil pipelines in the 1950s opened the way for the development of Botlek and Europoort. The introduction of the shipping container just 50 years ago has brought about the construction of two spectacular Maasvlakte port areas on land reclaimed from the North Sea. The information revolution of the last quarter-century has enabled the perfection of data exchange and has rendered the traffic of port and shipment information all but invisible, but it is a nonetheless crucial network for that.

It might take a little effort at first to view the port as a landscape in its own right. Back in the 1970s, we tended to speed through the Ruhr on car journeys as quickly as possible, windows tightly wound up. And that was how

NOL
MAERSK
MAERSK
MAERSK
MAERSK
COSCO
COSCO
COSCO

the industrial landscape of the Port of Rotterdam was also regarded in those days: as a necessary evil, dirty, faceless, not a place one would want to hang around in, excepting the dockers and the odd eccentric who made a deliberate choice to spend time there. In our primly-styled Western world, we are well familiar with and appreciate our housing landscape, and perhaps with the farming landscape; and we cherish the snippets of natural landscape that survive in the bits left over. The industrial landscape of a port is one that less readily lends itself to our aesthetic sensibilities. It is an acquired taste, this beauty; you have to learn to read this landscape. More than in any nature reserve one could imagine, this is a territory that feels like a wilderness in which you could get lost.

THE WELCOMING PORT

The historical 'hard' port has been moving inexorably westwards for decades, far beyond the city limits. Here, the old credo is still striven for: deeper berths, larger ships and more bulked transport. The spatial features here are large-scale, functional and serve their purpose, yet even in this industrial port landscape there is a tendency towards developing a welcome, albeit in an idiosyncratic way. Without doing away with the basic functionality of the port, the Port of Rotterdam Authority (*Havenbedrijf Rotterdam*) has been investing for nearly 25 years now in the spatial quality of the working environment. There has been a Quality Team in place since 2007, working on spatial transformations in the port, such as new beaches, roads, bridges, transport infrastructure and warehouses. The team is chaired by the Director of the Authority. Such a Quality Team, which just 30 years ago would have been written off as a bit of cultural dabbling, is helping shape the port landscape in our changed society. An attractive business location is a crucial factor to retain innovative businesses and joint ventures. Social acceptance goes hand in hand with physical quality, just as it does in the major national Dutch projects such as the Zuiderzee polders, the Delta Works and Schiphol Airport. A new kind of engineering thinking has taken hold in this port in the last few decades, one that fuses the port economy, nature and the environment, culture and environmental quality. How that is affecting the port landscape is something you can read about in our interview with Adriaan Geuze, one of the Quality Team leaders.

The port landscape is a remarkable mixture of functions guided by planning rules and the creative use of space. Peter Paul Witsen illustrates some of these fascinating locations. And Frank de Kruif explains just what exactly happens in the container port, the LNG port, the fruit port or the chemical port: What goods flow through this landscape 24 hours a day, and how is that trade conducted?

NEW ECONOMIC ACTIVITIES IN OLD PORTS

Nowadays, all the old urban ports worldwide have been taken over by city dwellers and in many cases been reworked into attractive mixed-purpose residential and commercial zones. In Rotterdam, the transformations of the Lloyd Pier, the RDM premises, Katendrecht and the Kop van Zuid have contributed significantly to forming Rotterdam's new image as a rough-and-ready yet welcoming metropolis. In the eyes of residents and visitors, the former port grounds lend the city a certain depth and resonance. The 'new economy' is coming to life in these former harbours in the city centre. There, companies are combining technology with creativity; artists and inventors are taking over the abandoned quaysides; and ship construction is returning too, but now in the form of luxury yachts and vessels to help dismantle drilling platforms rather than to build them. Many a wedding finds its backdrop in one of the restaurants along the old port jetties. The old port facilities have become rallying-points for a new kind of urban dweller, a kind that was seldom seen in the Rotterdam of the past. The world is discovering that there are more sides to Rotterdam than just the in-your-face industrial aspect. Rotterdam's new-found self-confidence is fuelled by the flair that has recently come to the old port. Based on 12 tendencies, Isabelle Vries shows what is going on, and which challenges the port economy will be facing.

EAT FRESH FROM THE VEGETABLE GARDEN

At urban farm 'Uit Je Eigen Stad' (From your own city), located in Rotterdam's Merwe-Vierhavengebied, you can eat fresh food grown in the vegetable garden. A former rail yard for the port has been replaced by free-range chickens and urban fields.

A DYNAMIC 'HULL'

My two-part essay Estimated Time of Arrival analyses the chronology of the Rotterdam port landscape and interprets it for the senses, drawing upon often very personal accounts of those who populate this landscape or indeed who were obliged to leave it behind. Captains of port industry such as the illustrious Swarttouw brothers and Cornelis Verolme have given way here to new entrepreneurs with new business models. Through interviews with a good number of port professionals, it has been possible to map the long historical lines of the port's development across an analysis of the key moments of change. The infrastructure of the berths and quays, railways and underground pipelines around the port – in other words, the port hardware – form, as it were, a dynamic hull to contain the intellect of the hundreds of companies that are based here and that can be seen as the port software. Design agency Beukers Scholma visualizes the 'facts and figures' of about 100 years of port construction and port trade. It can hardly be comprehended how many land surveyors, dredgers, civil engineers, electricians and shipbuilders have served their working lives in this port, and that is even before we consider the shipbrokers, shipping line owners, captains, pilots, dockers, boatmen and lashers. We then still have to put together the jigsaw of how all these elements, complex enough in their own right, work in harmony: from shipping movements to emissions, and from cargo manifests to customs declarations. It cannot be seen externally, but this huge assembly of firmware, to continue the computer idiom, is the port's true asset. Peter de Langen compares Rotterdam to eight other world ports, and shows what we can learn from them.

THE TREND FROM BRAWN TO BRAINS

While the port does still provide plenty of employment, the visibility of all those jobs is undergoing change. Two factors have caused this. The city and the working port have been driven apart, with the gigantic areas of imposing machinery now situated far to the west of town, partially offshore. Automation of port functions, which has been ongoing for several decades, is another reason why fewer and fewer people are required in the port. The *Polygoon* newsreels that used to give us details on how one ship after the other had been unloaded with muscle power have been replaced by promotional films showing completely automated mega-cranes unloading shipping containers from monster ships. A couple of employees now sit at computers in a hi-tech control room to guide the cranes. The dwindling of the physical labour force has been accompanied by an increase in the brain work performed in the port: the city of Rotterdam itself is number eight on the list of cities in terms of how many quality suppliers and port-related services they host (just behind Paris, which is the service centre for the port of Le Havre). Another palpable development is the tourist aspect of the port: the many cruise ships that anchor at the Cruise Terminal on Kop van Zuid each year. Once a hub of emigration, now it is where ships moor to let visitors in: the new travelling middle class of India, the United States and Brazil. Using the archives of the Port of Rotterdam Authority, Lara Voerman shows how the image of the port has been carefully constructed since the 1930s, and how this image has been adapted to keep step with changes in society.

City and port are working more closely together than at any time in the past, tackling a double-headed challenge. The city is keen to preserve a port atmosphere downtown and not to let these areas gentrify too much, and the Port Authority is seeking to bring urban life in abundance to the working port, without being untrue to the scale and dimensions of the port industrial complex. Both of these actors are well aware that the former and current port areas' strength is their uniqueness, and that buildings and infrastructure tend to grow in scale over time. The photography by Jannes Linders and Siebe Swart in this book shows both the city and the port, and invites the viewer to explore this sublime landscape. The new bond that has arisen between port and city is one that calls for an intricate balance between the two, not only in economic and planning terms but also in terms of how the city presents itself to the world. Rotterdam would lose its character if the ways of landlubbers ended up choking out the seafaring ways.

MINIATURE CITY

The *Oasis of the Seas* is by far the largest cruise ship in the world. It is a miniature city, with an on-board surf simulator, an ice rink, a carousel, a 'Central Park' and 25 restaurants. In October 2014, this 360-m ship docked at the Kop van Zuid's Wilhelmina Pier, after which it went in for maintenance at the mammoth dock of the Keppel Verolme shipyard.

HOLLAND-AMERICA LINE

In the nineteenth and twentieth centuries, America was a popular emigration country. While food shortages prevailed in some parts of Europe, the American government provided farmland virtually for free. Most of the Dutch emigrants left from Rotterdam, on the Holland America Line. This company was founded in 1873 by two young Rotterdammers named Plate and Reuchlin; Reuchlin drowned in 1912 with the sinking of the *RMS Titanic*. A crossing cost 30 guilders, which amounted to between one and three months' salary for a worker in those days.

HOLLAND AMERIKA LIJN
HOLLAND AMERIKA LIJN
N.A.S.M.

THE PORT IN NUMBERS

TOP TEN PORTS

BASED ON ANNUAL CARGO TONNAGE (2014)

source: Port of Rotterdam Authority 2015

Europe		World	
1	Rotterdam (NL) 444.7 million tonne	1	Ningbo/Zhoushan (CN) 873 million tonne
2	Antwerpen (B) 199 million tonne	2	Shanghai (CN) 755.3 million tonne
3	Hamburg (D) 145.7 million tonne	3	Singapore (SG) 580.8 million tonne
4	Novorossiysk (RU) 122.3 million tonne*	4	Tianjin (CN) 540 million tonne
5	Amsterdam (NL) 97.8 million tonne	5	Tangshan (CN) 500.8 million tonne
6	Algeciras (E) 95 million tonne	6	Guangzhou (CN) 500.4 million tonne
7	Marseille (F) 78.5 million tonne	7	Qingdao (CN) 480 million tonne
8	Bremerhaven (D) 78.3 million tonne	8	Rotterdam (NL) 444.7 million tonne
9	Ust-Luga (RU) 57.7 million tonne	9	Dalian (CN) 420 million tonne
10	Valencia (E) 67 million tonne	10	Port Hedland (AU) 372.4 million tonne

* Including the Caspian pipeline Consortium Marine Terminal

9 Ust-Luga
Bremerhaven 8
Amsterdam 5
3 Hamburg
Rotterdam 1/8
2 Antwerp
4 Novorossiysk
7 Marseille
Valencia 10
Algeciras 6
Tangshan 5
9 Dalian
Tianjin 4
Qingdao 7
Shanghai 2
Ningbo/Zhoushan 1
Guangzhou 6
Singapore 3
Port Hedlan 10

DEVELOPMENT

PORT CITY ROTTERDAM

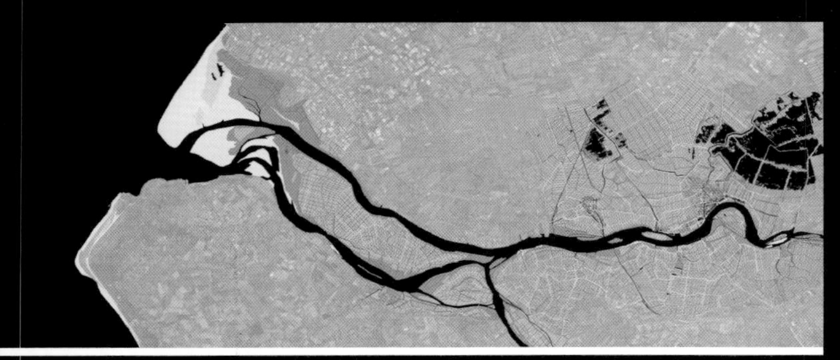

1850 90,000

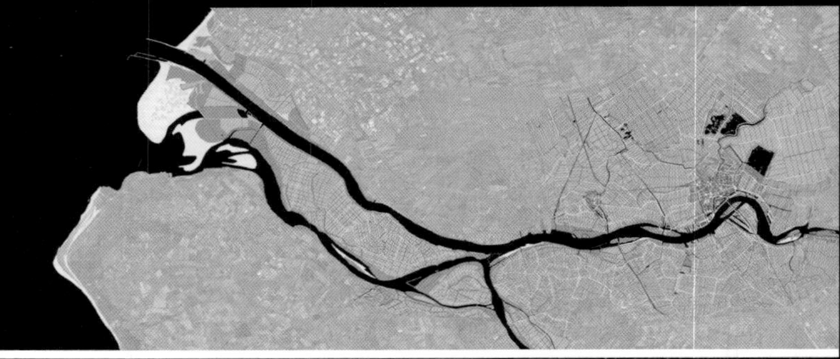

1890 201,800

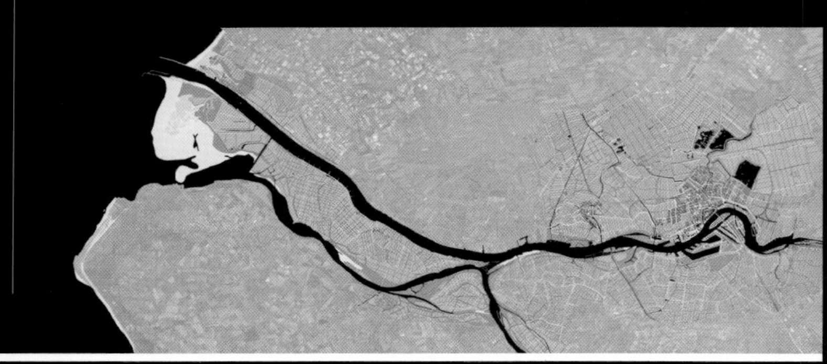

1907 403,300

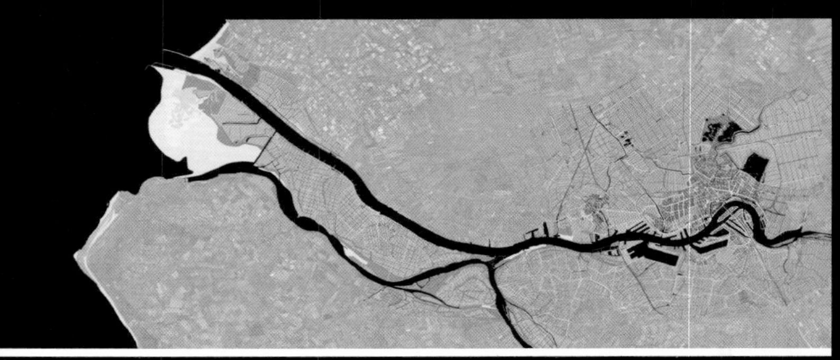

1920 506,000

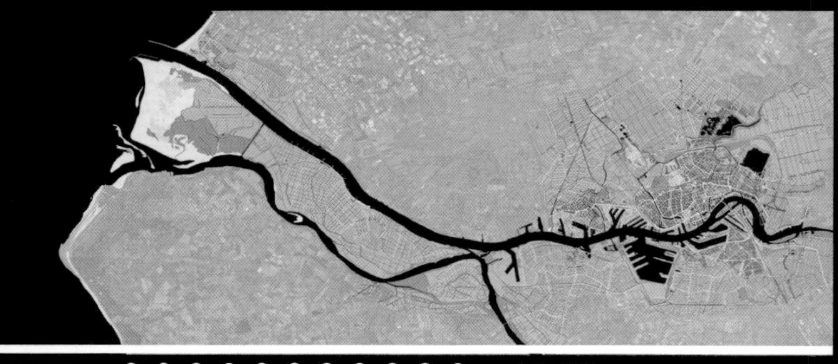

1937 595,400

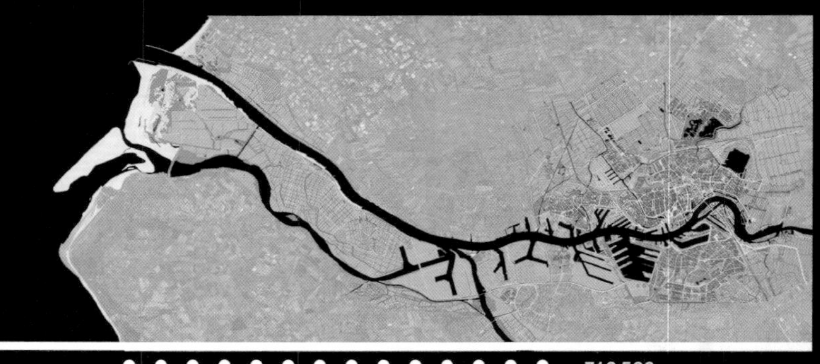

712,500

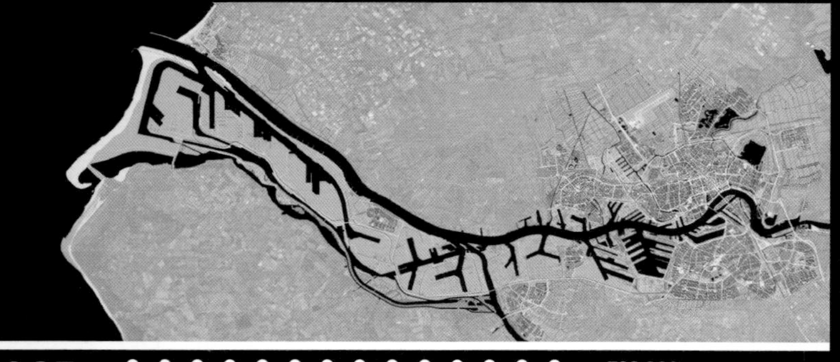

1965 728,300

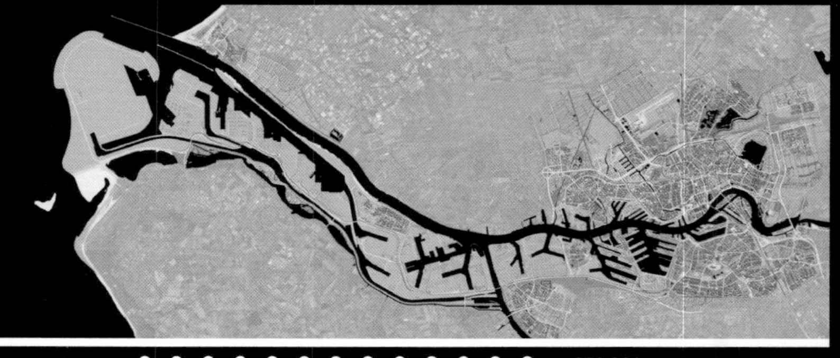

670,000

Growth of the port city and the population 1850-2015

source: 1850-2008: Mapping History / 2015: CBS / Port of Rotterdam Authority 2015

= 50,000 inhabitants

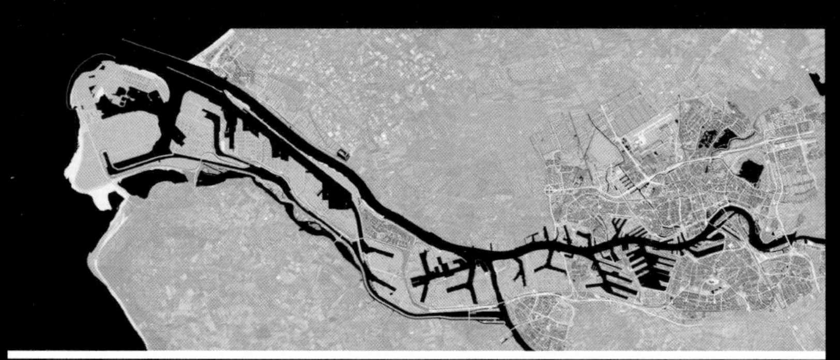

1980 568,100

1990 579,100

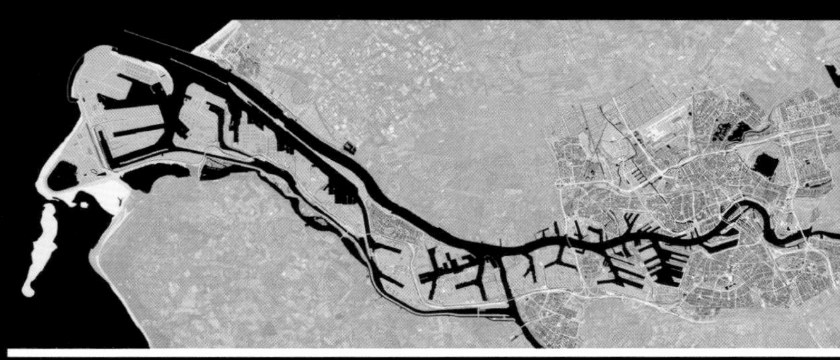

2000 592,600

2008 582,900

2015 623,652

SEA VESSELS

THE PORT OF ROTTERDAM

Avarage cargo tonnage per vessel 1900-2014

source: Port of Rotterdam Authority / PIM / CBL 2015

= 10,000 tonne

Avarage load per ship

1900

1930

1960

1990

2014

Avarage cargo tonnage from 1900 in proportion to a ship from 2014

source: Port of Rotterdam Authority 2015

Number of nautical accidents in 2014

source: Port of Rotterdam Authority 2015

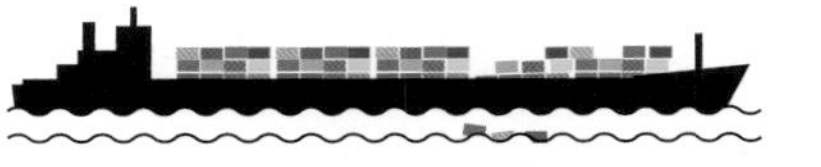

102 out of 27,790 visits

Ships visiting the Port of Rotterdam 1900-2014

bron: Port of Rotterdam Authority / PIM / CBL 2015

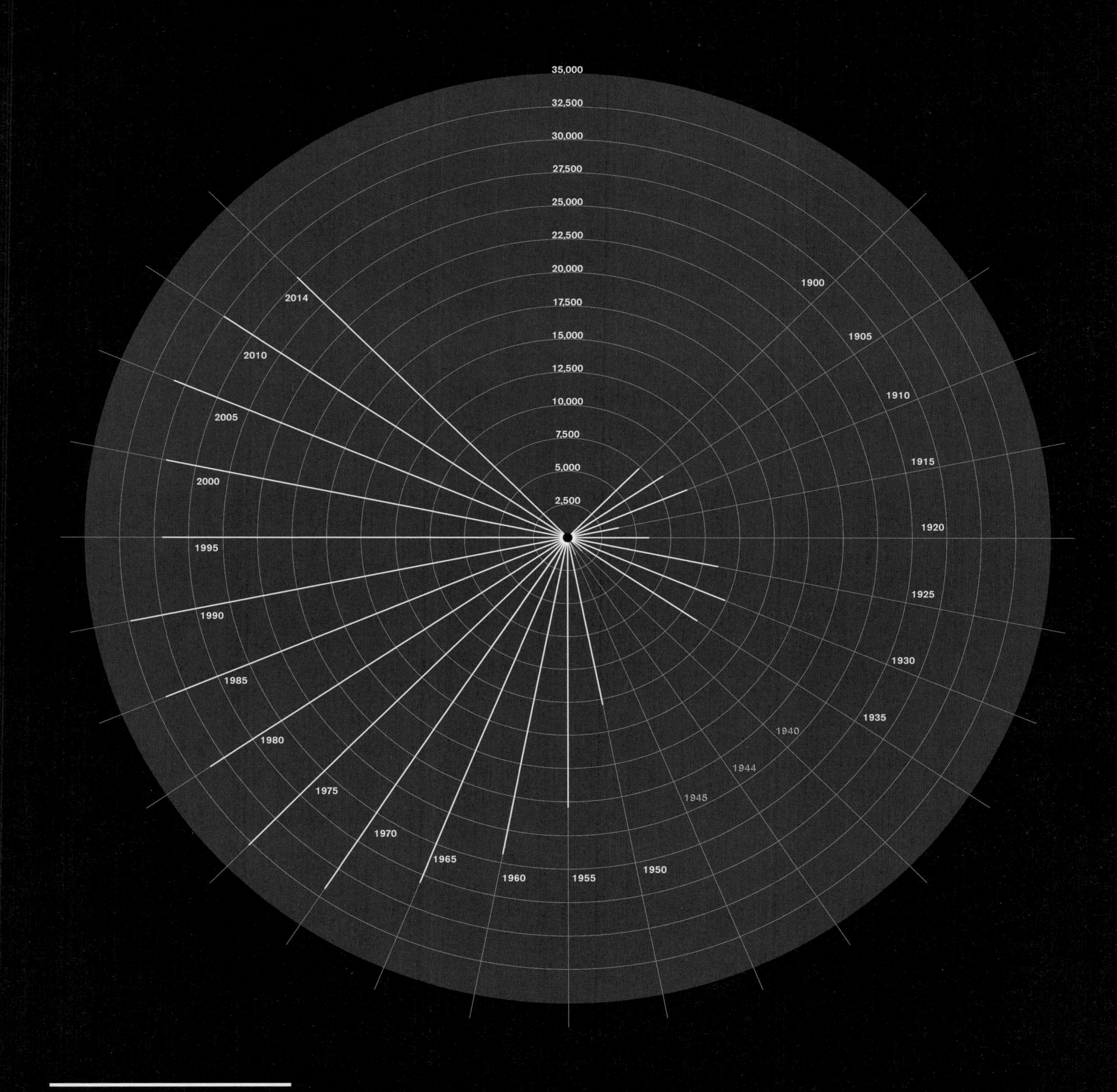

Division of vessels visiting the Port of Rotterdam in 2014

source: CBS 2015

ORIGIN / DESTINATION

PORT OF ROTTERDAM GLOBAL SEA TRANSPORT

Largest global transhipment through Rotterdam: petroleum and oil products

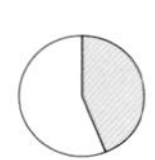
portion of incoming cargo
137 million tonne

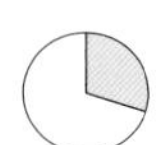
the majority comming from
Russia with 41 million tonne

Seccond place in global transhipment through Rotterdam: containers

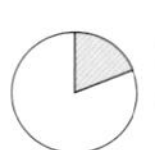
portion of incoming cargo
60 million tonne

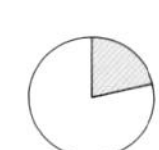
the majority comming from
China with 13 million tonne

Third place in global transhipment through Rotterdam: ores

portion of incoming cargo
33 million tonne

the majority comming from
Brazil with 22 million tonne

Largest global shipment from Rotterdam: oil proucts

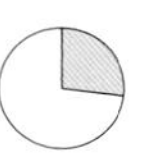
portion of outgoing cargo
35 million tonne

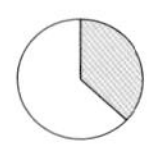
the majority comming from
Singapore with 13 million tonne

TRANSPORT AND INDUSTRY

PORT OF ROTTERDAM EMPLOYMENT AND TRANSHIPMENT

Direct port-related employment 2013

source: Erasmus Universiteit Rotterdam 2014

= 500 employees

total employees
93,766

Transport

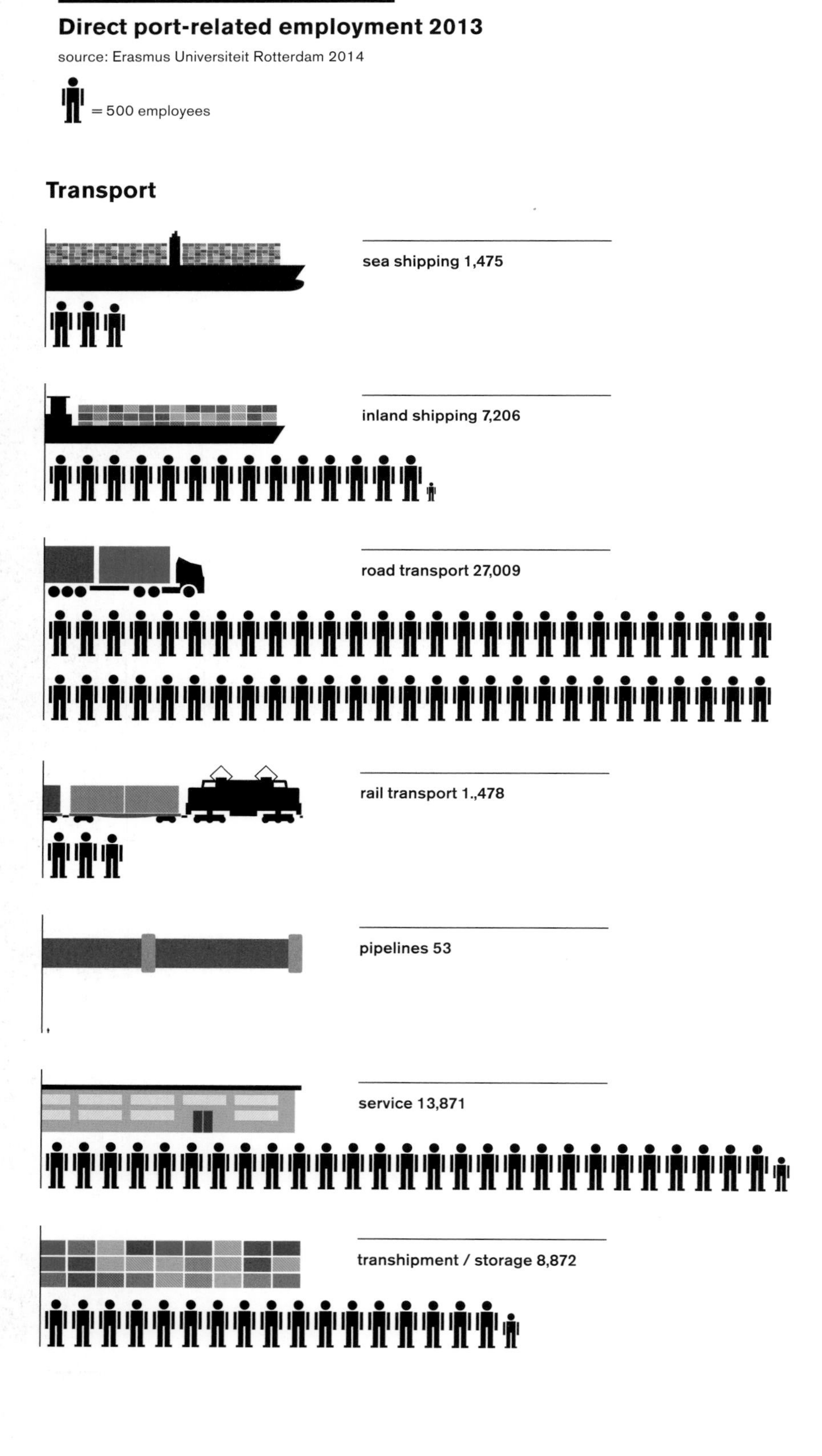

Industry

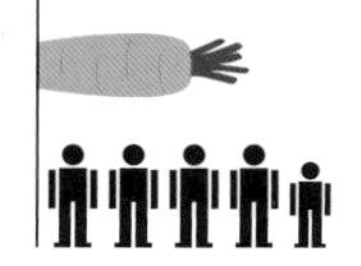

food 2,417

petroleum 3,363

chemical 4,740

base metals / metal procucts 3,317

transport 1,963

electricity generation 1,874

other industries 2,748

wholesale 7,982

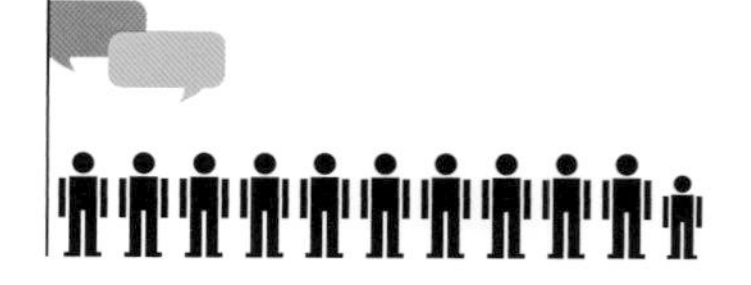

business / non-business services 5,398

Transhipment of goods in tonnage 1900-2014

source: Port of Rotterdam Authority / PIM / CBL 2015

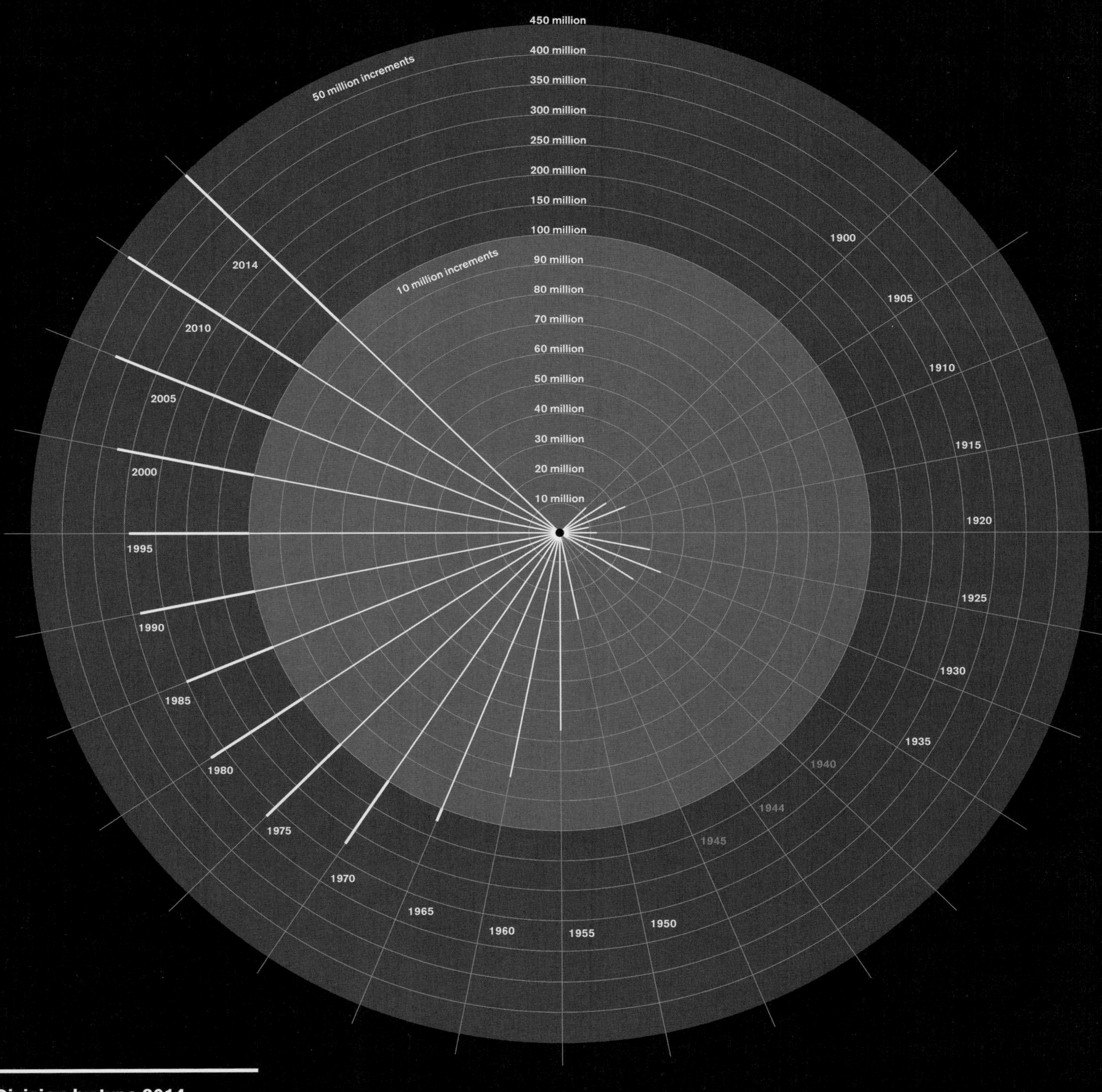

Division by type 2014

source: Port of Rotterdam Authority 2015

incoming

outgoing

INFRASTRUCTURE

LOGISTICS OF THE PORT OF ROTTERDAM

Total size and length 2014

source: Port of Rotterdam Authority 2015

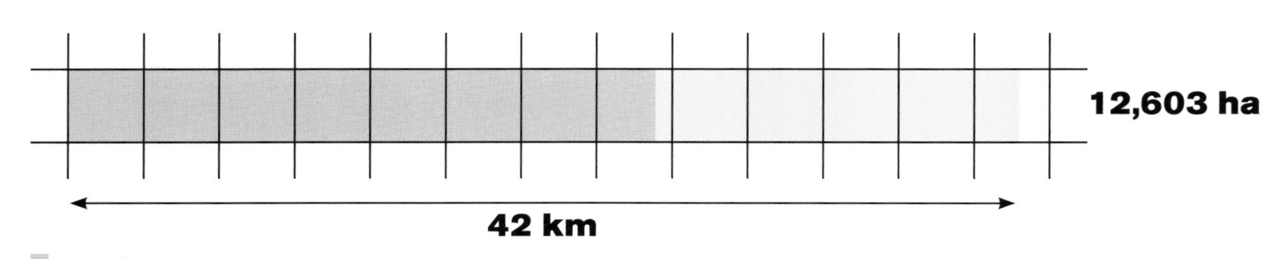

land
water

Maximal water depth (NAP) Eurogeul in the North Sea 2014

source: Port of Rotterdam Authority 2015

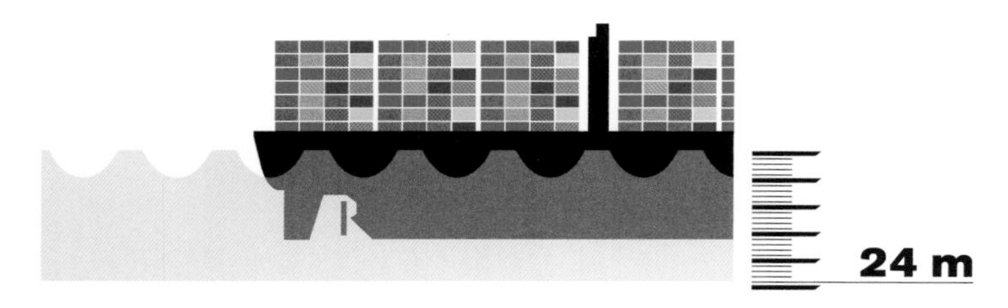

Quaysides and banks 2014

source: Port of Rotterdam Authority 2015

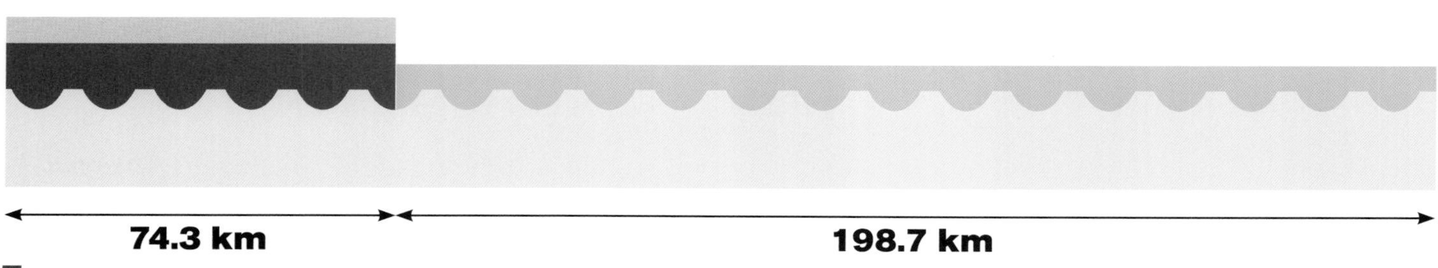

quaysides
banks

Sites and terminals 2014

source: Port of Rotterdam Authority 2015

oil refineries, refinery terminals, independent tank terminals for oil products

chemical plants, biofuel factories, vegetable oil refining plants

container terminals: deep-sea, short sea, empty depots

general cargo: roll-on/roll-off, other

bulk dry cargo terminals: agricultural commodities, ore, coal, biomass, etcetera

Pipeline network from the port 2015

source: Port of Rotterdam Authority 2015

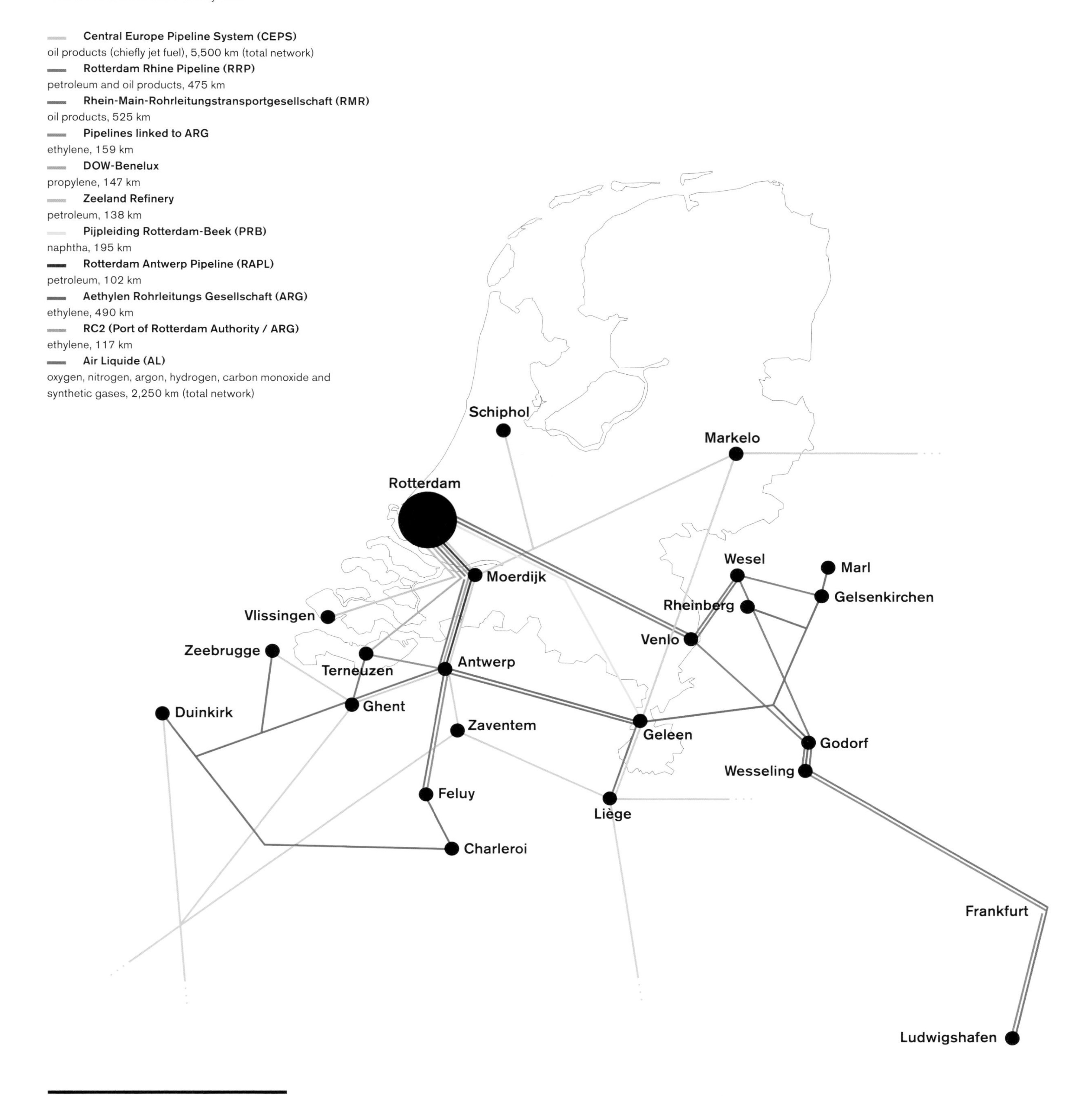

Tank storage in the port 2014

source: Port of Rotterdam Authority 2015

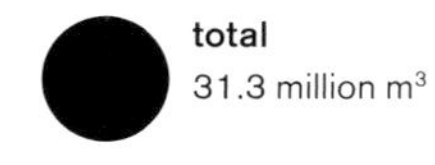

total
31.3 million m³

petroleum
14.5 million m³

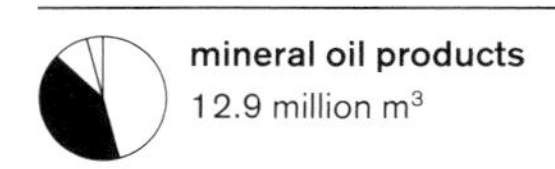

mineral oil products
12.9 million m³

petro-chemical oil products
2.7 million m³

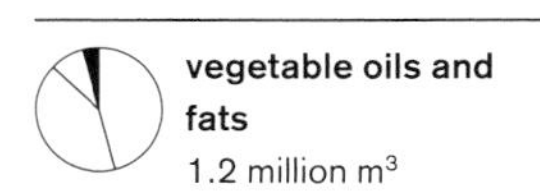

vegetable oils and fats
1.2 million m³

CONTAINERS

THE PORT OF ROTTERDAM

Container ship in 1967 and 2015

source: Maritiem Museum Rotterdam / Port of Rotterdam Authority 2015
A TEU (Twenty-foot Equivalent Unit) is 6.1 m long, 2.44 m wide and 2.59 m tall

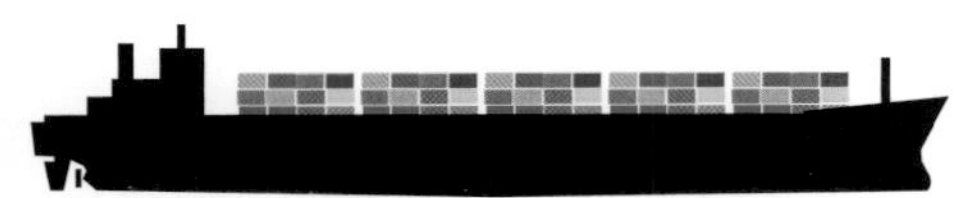

first container ship that used the ECT in Rotterdam in 1967
MSC Atlantic Span, 197 m, 596 TEU capacity

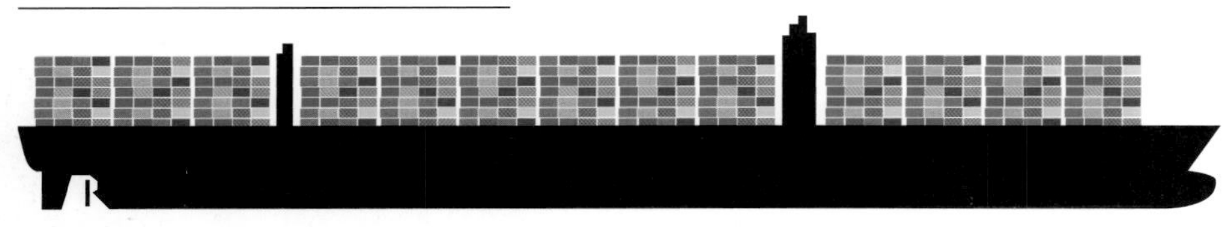

largest container ship in the Port of Rotterdam in 2015
MSC Oscar, 395.4 m, 19.224 TEU capacity

19,224 containers

Length *MSC Oscar* = 2x *Euromast*

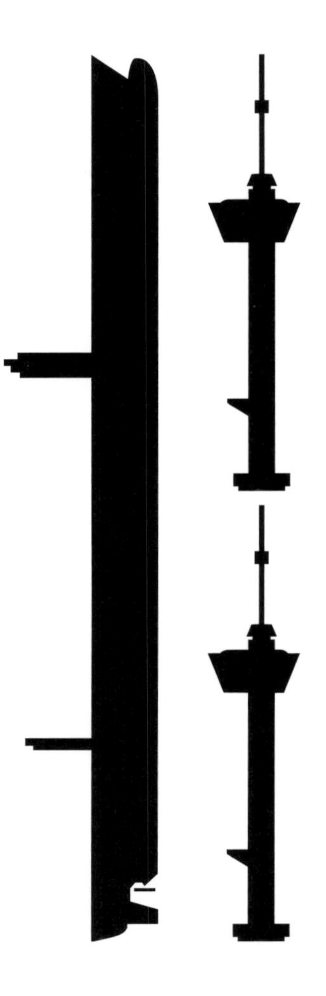

Transporting the containers from the *MSC Oscar* to the hinterland

source: Port of Rotterdam Authority

19,224 containers via road transport

8,112 articulated lorries with trailers needed

19,224 containers via inland shipping

92.5 barges with a capacity of 208 TEU each needed

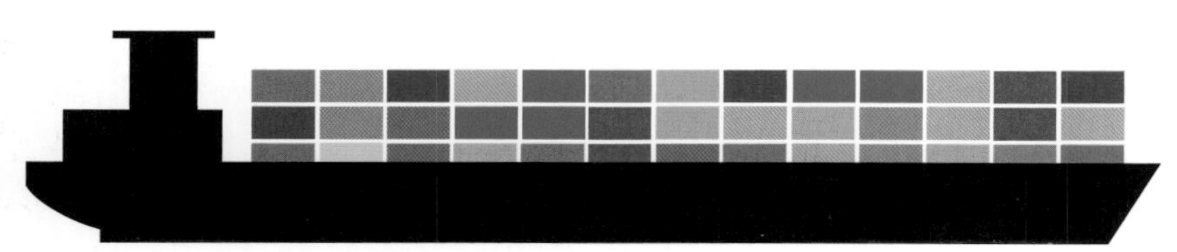

19,224 containers via rail

246.5 trains with 39 carriages each needed

Transport to the hinterland 2014

source: Port of Rotterdam Authority

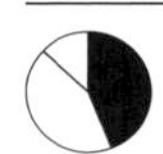

portion lorries

portion barges

portion trains

Number of incoming and outgoing containers from the Port of Rotterdam 1970-2014

source: Port of Rotterdam Authority 2015

incoming
outgoing

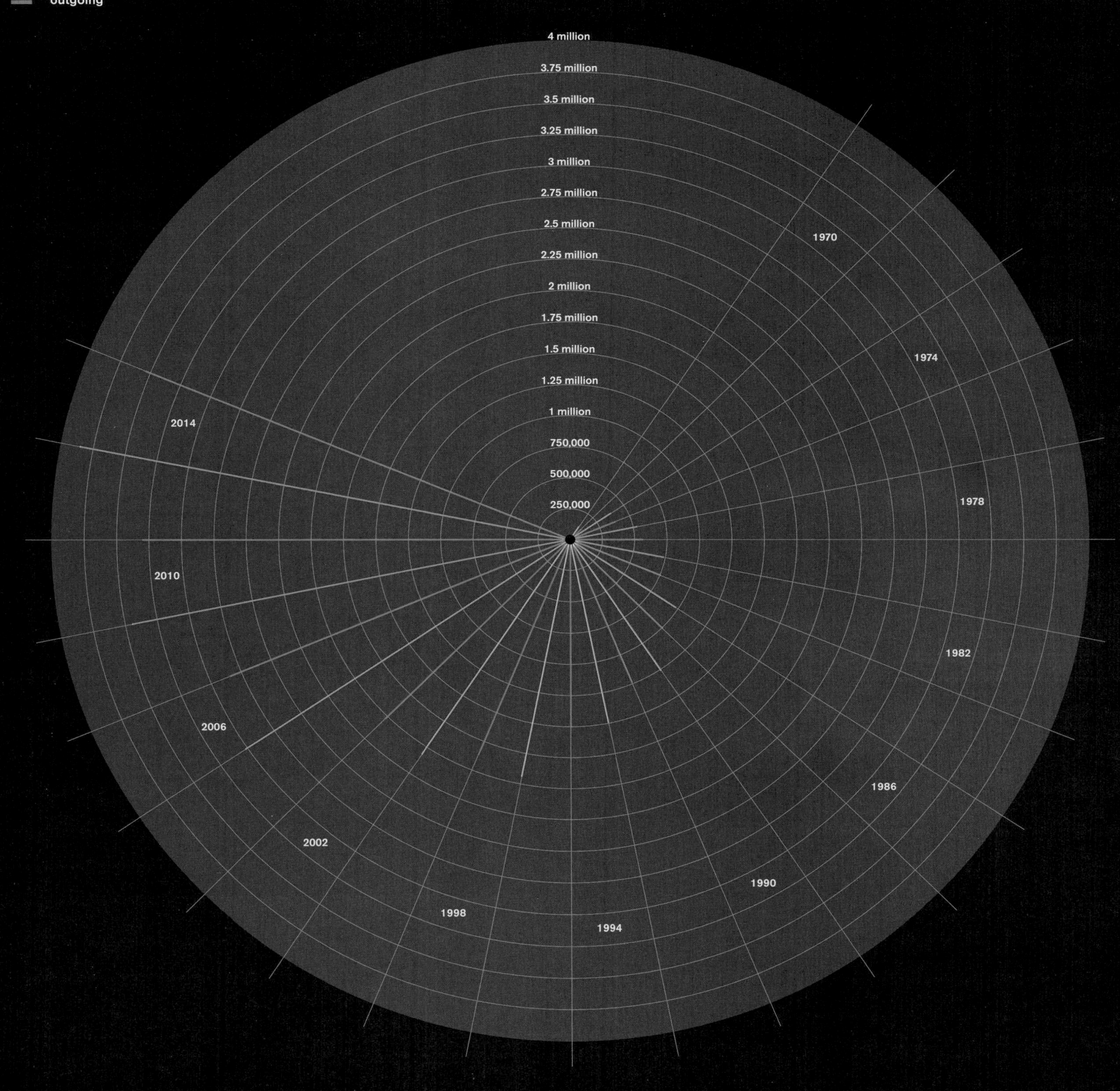

Top five European container ports 2014

source: Havenautoriteiten / Port of Rotterdam Authority 2015

Rotterdam

Hambu

Antwerp

Bremerhaven

Le Havre

PORT PLACES

Frank de Kruif FdK
Isabelle Vries IV
Peter Paul Witsen PPW

photography
Jannes Linders
Siebe Swart

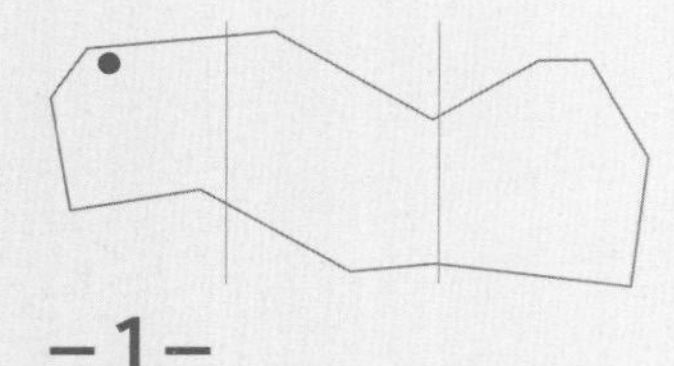

– 1 –

Prinsessenhavenweg \ Maasvlakte Rotterdam

MAASGEUL

Safely Entering the Port

Every year, about 30,000 ships enter the port. The crew sails the ship across the sea towards Rotterdam. But once the Hook of Holland comes into view, they need to seek outside assistance. Nautical service providers ensure that the ship winds up safely at a quay, jetty, post or buoy.

The first person that a captain will encounter is a registered pilot. He steps aboard, after being dropped off by a pilot boat or helicopter, to assist the captain in navigating the final steps to the port. This stretch is always the most difficult: tides, currents and shallows can be dangerous, especially if the captain does not know the port. Larger ships and ships with dangerous cargo are required by law to use a pilot, though advances in technology mean that the number of ships requiring a pilot is decreasing.

Once the ship has arrived in the port, tugboats take over the guidance procedure to bring the ship to its berth. In contrast to pilotage services, which are a legal monopoly, the port's towage services have to compete for their customers, namely the shipping companies. Well-known service providers such as Smit, Kotug, Fairplay and Svitzer operate not only in Rotterdam and other Dutch ports, but all over the world.

The final stage of the job is for the 'rowers'. They secure ships using ropes, usually attaching them to bollards on the quay. They used to do this from a rowboat, which explains their name; this name lives on in, for example, the name of the Roeiers Vereeniging Eendracht. But these days they have their own specialized work boat, called a punt.

In addition to nautical service providers, there are also many other specialists operating in the port, such as companies that supply ships with supplies or parts, or that perform maintenance on the ships. Although their activities differ, all of these service providers have one thing in common: in the 24-hour economy that is the Port of Rotterdam, they are on the job day and night, seven days a week. Rain or shine.

– FdK –

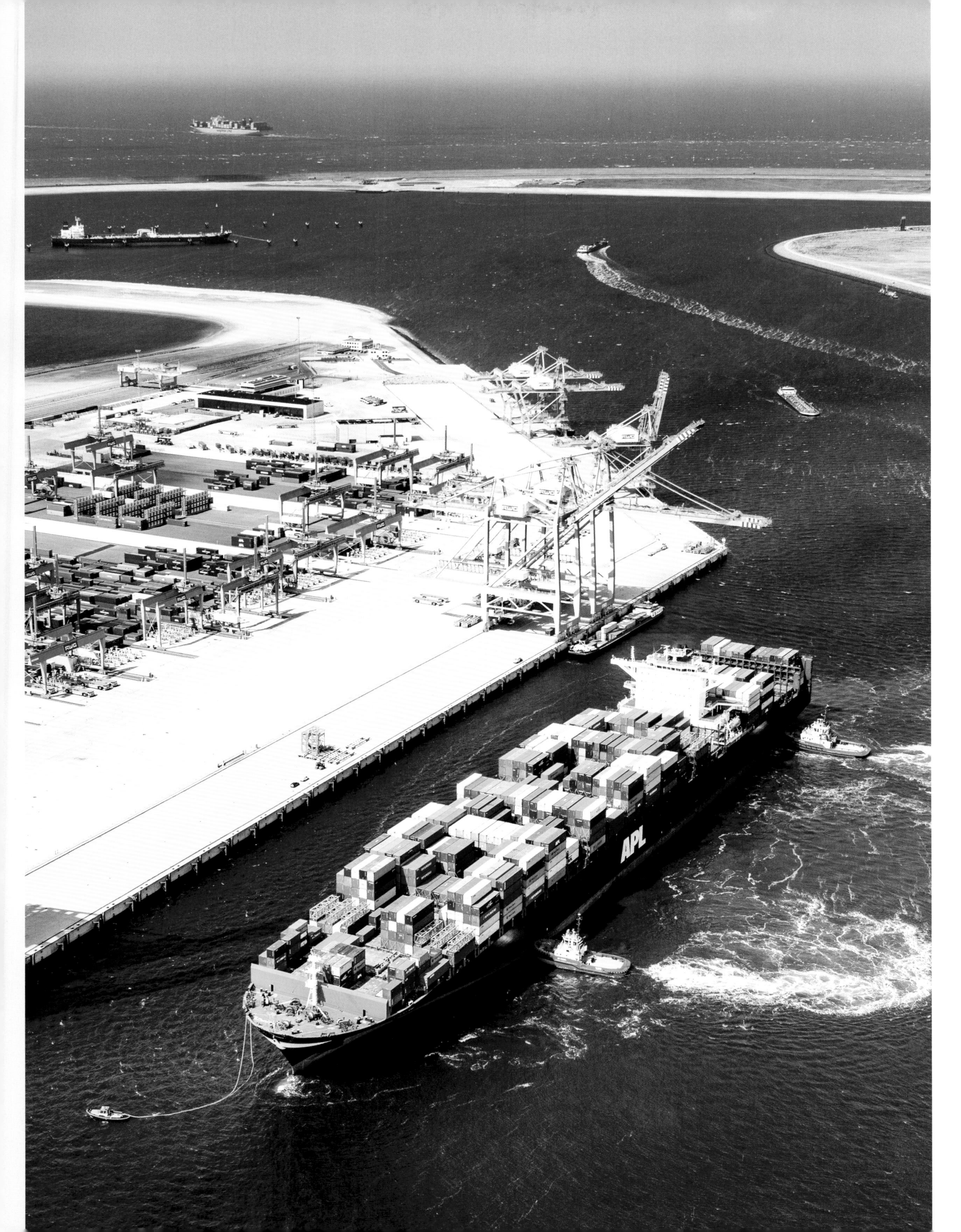
APL

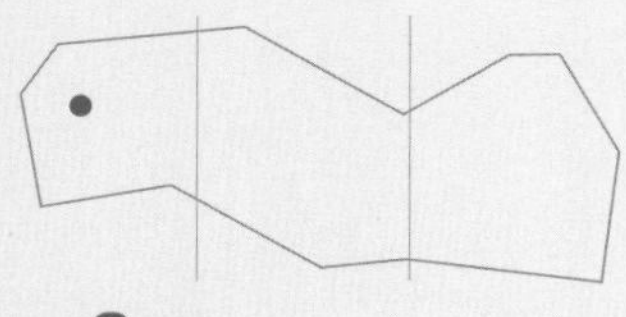

– 2 –

Europaweg 910 \ Maasvlakte 2 Rotterdam

PRINSES AMALIAHAVEN

Automation

The stevedore known as APM Terminals has the first zero-emissions terminal in the world. Across its entire transhipment site, there are no emissions of CO_2, NOx or particulate matter. APMT is located on the Maasvlakte 2. The unloading process is completely automatic, as is the stacking of the containers and the loading of the lorries. The fully automatic, unmanned vehicles run on batteries, instead of a roaring diesel engines. And the batteries are powered by green energy. Along the quays, the possibility of 'shore power' is built in, waiting for the moment that the first container can insert the plug into the socket.

A long time ago, developments in crane technology meant that the traditional form of manual labour, namely lugging, had disappeared from the port. And now this latest development means that the classic crane operator is also disappearing. At APMT, he no longer sits in the crane, but instead coordinates the whole process from the office. From his office window, the crane operator can control three cranes at the same time. Because they are unmanned, these cranes can move faster, even though they are almost as high as the Erasmus Bridge. This automation leads to savings in terms of time, energy, costs and emissions.

But automation in the port also has many implications for the labour market. It's not as if there will soon be no one working in the port. There will be more jobs for skilled personnel: programmers, controllers, process operators and green engineers. But a development such as the one at APMT has many implications for the traditional jobs in and around the terminal, for example rowers, who still do the traditional manual work.

These technological developments will not stop. Experiments are underway with self-driving vehicles, unmanned ships and ships that can moor automatically to magnetic docks. Companies such as Google are getting involved in the logistics sector. From A to Z, everything is under control. And all of this will continue to further drive progress in terms in safety, efficiency and costs.

Will the longshoreman become the miner of the twenty-first century? Hopefully not. But to prevent this, social innovation in the port is vital.

– IV –

MAERSK LIN

MAERSK
MAERSK SEALAND
P&O
109

KALMAR
KALMAR
53
stinis
stinis
137
124
127
119
106

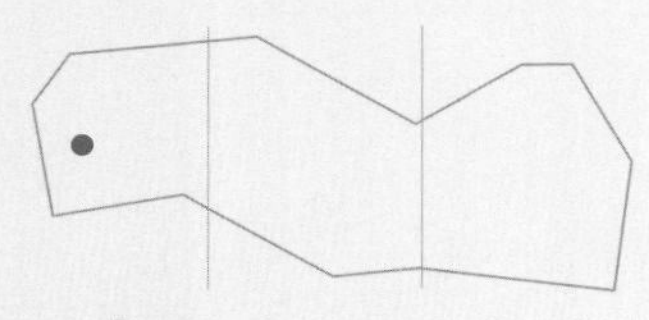

– 3 –

Noordzeeboulevard 501\ Maasvlakte Rotterdam

SLUFTER

Breeding Birds on Sludge

A platform floats on the water; it is intended as a breeding place for birds, namely common terns. But deep below the surface here are heavy metals that have accumulated over the years along with the dredging sludge from the port. This is the Slufter, a deep pit on the Maasvlakte surrounded by a 24-m ring dike made of sand covered in a layer of clay, designed to store contaminated dredging spoil.

When the Slufter opened in 1987, the expectation was that it would be full by 2003. At the time, three quarters of the dredged sludge was contaminated. Industries further upstream, deep into other countries, were discharging their waste water directly into the Rhine and the Maas. But this hardly happens any more, as industrial production and purification techniques have greatly improved. Contaminated dredge now comes mainly from older harbour basins that were contaminated in the past. This amounts to only about 10 per cent of the dredged sediment.

Sludge and sand settle on the waterbed, making the water shallower. For each harbour basin and each channel, the Port Authority guarantees a certain water depth. The point is that the keel of a ship sometimes has no more than half a metre of clearance. Pilot boats regularly measure the actual depths. If the measured water depth approaches the minimum level, a 'trailing suction hopper dredger' comes along to vacuum away a layer of sludge sediment.

The Port Authority also determines the quality of all the waterbeds every year. Before a dredger starts to work, he already knows whether the sludge that he will bring up is clean or polluted. If it is clean, he takes it straight to the North Sea. If it is too polluted for that, it goes to the Slufter. That still amounts to about 500,000 m^3 per year, enough to fill the entire Ahoy stadium. But the Slufter was designed to be able to hold much more than that. The well is now half full.

The Port Authority is now investigating different opportunities for using the water surface. There has been a trial with growing algae as a raw material for organic plastic, and perhaps solar panels will float here in the future. Common terns had already discovered the opportunities of the Slufter, and began to nest on the edges of the depot. But what the birds did not notice was that the water level rises whenever sludge is deposited, and that meant that their egg clutches were washed away. That's why another solution was eventually found. It may look unnatural, even industrial, but for the common terns it is perfect: a Styrofoam platform that bobs on the surface, surrounded by a fence, with many short tubes that offer protection from birds of prey.

– PPW –

The area around the Slufter is used for recreational purposes as will a disposal site for batteries, which are buried din the hill.

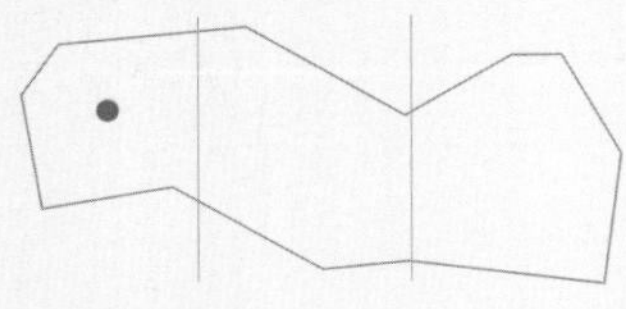

— 4 —

Europaweg 875 \ Maasvlakte Rotterdam

AMAZONEHAVEN

A Giant Step in Containerization

The ECT Delta Terminal at the Maasvlakte. The largest container vessels berth here. One record-breaking ship after another arrives. And this has always been the case, since the opening of the Delta Terminal in the mid-1980s. And even before then, the ships arrived at the Beatrixhaven, which was later renamed the Home Terminal. The first containership that moored there, in 1967, was capable of carrying 596 TEU (Twenty-foot Equivalent Unit). The latest record-breaking ship that entered the Delta was in early 2015, with a capacity of 19,224 TEU. This is the kind of scale increase that has occurred over the past 50 years in the field of container shipping.

Back then, the container was nothing less than a revolution, in several respects. Container shipping made the global economy possible, changed the how things were done in ports and changed the face of the ports themselves. The container was a huge step forward in the efficient transportation of goods. Consignors could now load their cargo on their own premises, seal the container, send it to the port via lorry or train, load the container onto a ship at the port and then have it unloaded somewhere else on the other side of the world, and finally be delivered to the recipient.

The result was that the cargo soon became invisible, even on the quays in the port. What you see today are the neatly stacked multicolour boxes. When you fly over them, they look like Lego bricks.

More and more goods – especially the traditional breakbulk cargo – have disappeared into containers over the years. With the advent of the refrigerated container, that also happened to fresh products such as fruit and meat. Anything that can fit in a container will be put in a container. Except if it is an urgent job, in which case the shippers send their goods by air.

With initially only the Home Terminal, and later the Delta Terminal and much later the Euromax Terminal, ECT has long been the dominant supplier of container transhipment in the Port of Rotterdam. There used to be smaller competitors such as Hanno and Uniport, but they lacked the scale to be able to pioneer in the automation of transhipment, which ECT Delta managed to do. The second major terminal on the Maasvlakte, the APM terminal, mainly handled the ships of its sister company Maersk Line. With the arrival of two new terminals on the Maasvlakte – one an extension for APM, the other a new terminal for Rotterdam Word Gateway (RWG) – the container market will continue to grow, but there will also be more competition.

Despite the fact that the outsourcing of production to low-wage countries seems to be in decline, while a development such as 3D printing is on the rise, at the moment there seems to be no end in sight to the increase in container transport. And that means a growing market for all terminals in the port.

– FdK –

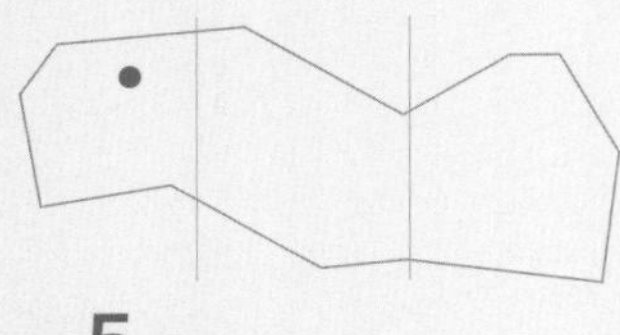

– 5 –

Maasvlakte Rotterdam

SLAG MAASMOND

Size and Scale

On Slag Maasmond, with the view to the north, you stand ringside to greet one of the world's largest ships. On 3 March 2015 the MSC Oscar, at that time the largest container vessel in the world, moored in Rotterdam. The ship can carry over 19,000 Twenty-foot Equivalent Unit (TEU) containers. Almost ten years earlier, Rotterdam celebrated the arrival of another record-breaking ship, the *Emma Maersk*, with 14,000 TEU on board. To illustrate what that really means: the containers of the *MSC Oscar* would form a line of 115 km if stacked end to end, the distance from the Maasvlakte to Utrecht. We will allow that to suffice in terms of explaining the numbers.

The history of the port shows that there is one trend that still seems to apply: scale. This phenomenon has to a large extent determined the face of the port. Engineer Gerrit Jan de Jongh knew what Rotterdam needed to become the best port in the world. You cannot deal with space in a childish way. De Jongh was the design genius behind the major port expansions around 1900: the Rijnhaven, the Maashaven and his pièce de résistance, the Waalhaven. This Director of Public Works knew nothing about the container, because that box, which unleashed a revolution in global trade, was not used on the quays of Rotterdam until 1966. Yet the Waalhaven still operates today as an excellent port area, albeit in a very different form than the designer ever envisioned. That is the art of design: to make something that can keep up with the times and that can adapt. The unpredictability of some changes forces us to think in terms of flexible concepts.

The giant container ships also represent other challenges, in addition to design issues. How, for example, do you receive such a mammoth craft safely, and remove some 10,000 boxes quickly? Are there enough rowers, tugs, lashers and cranes available? When can the inland barges be called? Can the underlying infrastructure handle it all?

The main incentive behind scaling up remains the cost savings, but there are also environmental benefits. In order to become even more efficient, shipping companies merge, or band together to form alliances. The stevedores at the port compete hard with each other. But they will increasingly have to work together to make optimal use of nautical services, hinterland connections and labour.

Will this scaling continue, or has the optimum already been reached? Vessels of 24,000 TEU are reportedly already on the drawing board, somewhere. Whoever knows where this trend will end can tell you the answer.

– IV –

GASOLINE
CONTAINS LEAD
NO SMOKING

COSCO

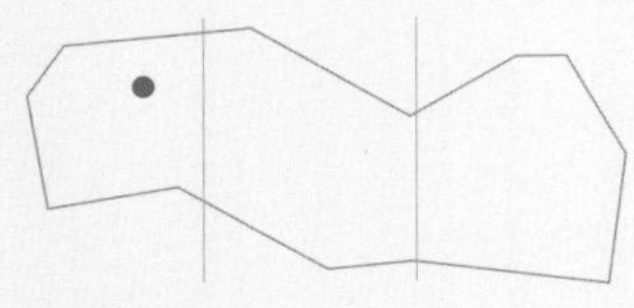

– 6 –

Maasvlakteweg 975 \ Botlek Rotterdam

8E PETROLEUMHAVEN

An Invisible Junction

Before the Maasvlakte 2 existed, the storage tanks of the MOT marked the end of the port. You could not go any further. And anyone who went there could be forgiven for not realizing that this piece of the port was perhaps the largest logistics hub in the Netherlands. And still is.

More than any other port in Europe, Rotterdam is driven by oil. For decades, the supply of crude oil has been fairly consistent, usually about 100 million tonnes per year, sometimes slightly more. That means that for a long time, the black gold was the port's most important type of good in terms in volume. Only since 2007 has Rotterdam stored more containerized cargo than oil.

Every year, about 240 tankers deliver about a third of that 100 million tonnes to the Maasvlakte Oil Terminal, which is the MOT's full name. It can handle even the biggest supertankers. But that's not what makes MOT special.

What is remarkable, firstly, is that the terminal is owned by all the major oil companies that have refineries in the port area, together with tank storage company Vopak. In the early 1970s, these companies predicted enormous growth in the delivery of oil, which meant that they urgently needed additional storage capacity. The only place that was eligible was the Maasvlakte, which was then still new and which had very deep waters. Because there was not enough space to give individual companies their own places, it was decided to build a joint facility. The second remarkable feature is the invisibility of the MOT as a hub. The transport of oil takes place underground: the refineries in the port are fed with oil via pipelines. But this underground distribution does not stop in Rotterdam; oil is pumped through these pipelines all the way to Vlissingen, Antwerp and the Ruhr Region.

This pipe network largely explains why Rotterdam is in such a strategically strong position in terms of Western European oil imports. And in terms of strategy, is it rumoured that part of the Netherlands' strategic oil reserves are located at the MOT. Should the supply ever stop, then the Netherlands could manage for a few months with oil from the MOT.

– FdK –

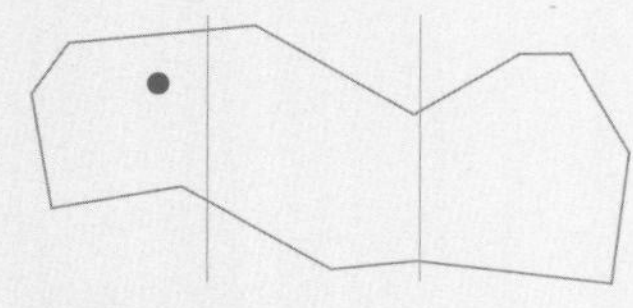

– 7 –

Maasvlakteweg 991 \ Maasvlakte Rotterdam

BEEREILAND

An Uninhabited Island to Sail Around

Near the access from the sea to the Maasvlakte is a small island that would appear to be dangerously in the way of the giant container ships that turn towards the port here. It is called the Beereiland, named after the nature reserve that was located here long ago. Until about 2010, the island was connected to the Maasvlakte, as part of the Papegaaienbek (Parrot's Beak), the northernmost jetty of the First Maasvlakte, so called because the curvature of the pier resembled the beak of a parrot. What was then a popular spot for fishermen is now the domain of seals and black-backed gulls.

Thanks to the soil structure and the drainage system, the Beereiland functions as an ecological *stepping stone.* But that is not why the island was constructed, or, more accurately, why the connection to the mainland was dug up. The reason was the 2011 arrival of the LNG terminal known as GateATE (Gas Access to Europe). LNG stands for Liquefied Natural Gas, which is natural gas that has been liquefied by extreme cooling, to 162 degrees below zero. LNG is emerging as a sustainable fuel for maritime and inland shipping. Its emits far fewer pollutants than conventional diesel fuel, and also takes up considerably less space than it does in gaseous form, which means it can be transported over long distances. The LNG in the GateATE terminal comes from countries such as Trinidad, Algeria and Australia.

Sustainable or not, LNG is explosive stuff. Much like with petrochemicals, its storage and logistics call for special security measures. The Beereiland is one of these measures. The water between the island and the terminal is exclusively meant for LNG carriers and the supply vessels that carry the liquid gas further into the port. All other ships sail outside of this water. If a vessel were to sail off course for whatever reason, this arrangement means that it would not crash into a tanker that would release flammable materials, but instead would safely run aground on an uninhabited island.

– PPW –

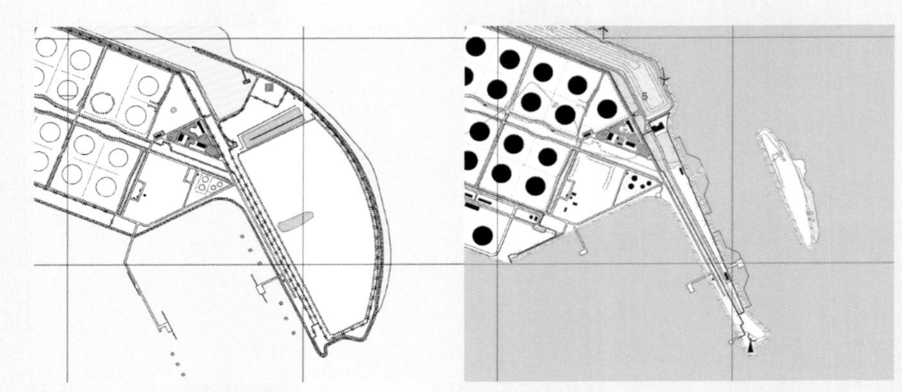

The Papegaaienbek (Parrot's Beak) before and after the excavation for the LNG terminal in 2011.

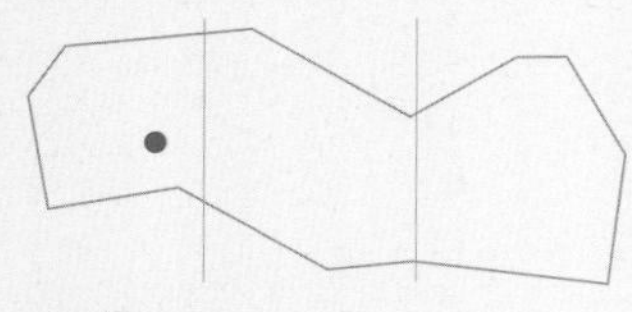

– 8 –

Noordoever Oostvoornse Meer 1 \ Oostvoorne

OOSTVOORNSE MEER

Salted Water

There are not many salty inland lakes in the world. They are only created when an inlet is cut off from the open sea. This is exactly what happened in 1966 at what is now the Oostvoornse Meer. Sand was needed for the construction of the First Maasvlakte. It was brought in from the waterbed of the Brielse Gat, which was the inlet that remained after the mouth of the Brielse Maas had been dammed in 1950. To be able to extract this sand, the seawater had to be tamed and a second dam was necessary.

The formation of the Oostvoornsemeer put a chain of subsequent effects into motion. The sheltered, salty water proved attractive to many plant and animal species that are not often found in the Netherlands, or even in Europe. The green beach, the swampy southern shore of the lake, was designated (as part of the Voornes Duin) a Natura 2000 area. Natura 2000 is the European network of valuable nature areas where strict planning requirements apply. No interventions are permitted that might affect the quality of the habitat for the protected plants and animals, including in the surrounding area. The Port Authority still has to take this into account in its plans for developing the Mississippihaven.

The new inland lake also turned out to be a popular recreation area, especially with divers. The water was clear, because the salt content prevented the growth of algae. There are many shipwrecks under the water, because there used to be a treacherous sandbar here. Divers and archaeologists were both drawn to this site.

In this way, the Rotterdam region acquired a unique natural and recreational lake within its borders. But the joy was not to last long. Rainwater and fresh groundwater mixed with the former seawater. The lake became a freshwater lake, and its water began to blur. The solution was found at the bottom of the Beerkanaal, between the Europoort and the Maasvlakte. Salt water accumulated in a well on the channel floor; this water is heavier than freshwater, so it sinks to the bottom. It was relatively clean, because the pollutants flowed along with the freshwater towards the North Sea.

Between the Beerkanaal and the Oostvoornsemeer, a pipeline was installed that pumped saltwater into the lake; this pipeline was completed in 2008. It turned out to be good for nature, good for divers and good for archaeologists. But shipworms also feel at home in this sheltered, salty water. And these molluscs like to bore into damp, old wood, for example shipwrecks.

– PPW –

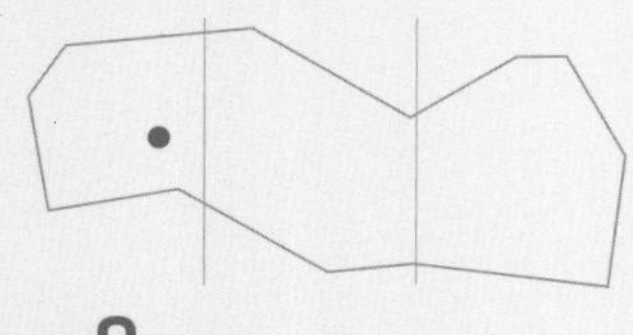

– 9 –

Beerweg \ Maasvlakte Rotterdam

KRABBETERREIN

Pioneering Toads and Orchids in the Pipeline Corridors

The port has long strips of land that will always remain green and undeveloped, the so-called pipeline corridors. Above ground, they are recognizable as grass fields amid the logistical and industrial activities. Under the ground here is an infrastructure that can compete with road, rail and inland waterways. The port contains no less than 1,500 km of pipeline (which is the distance from Rotterdam to Helsinki), bundled in corridors. Crude oil, chlorine, steam, electricity and much more are transported safely and reliably under the ground, sometimes from one company to another, and sometimes even into the German hinterland.

The grass on these pipeline corridors is kept short. If a calamity were to cause fluid from the pipeline to seep upwards, it might remain undetected for a long time if the grass were higher. Marking poles stick out above the grass, which allow the pipeline corridors to be recognizable above ground. They are dug up with some regularity, for example for maintenance work. Trees do not belong on these pipeline corridors, as their roots can damage the pipes. But in the springtime, what looks like bare grassland changes into a sea of purple orchids.

Along with other plants, the fen orchid has found a niche above these pipelines, for example at the Krabbeterrein. That is special, because this species is one of the rarest wild orchids in Europe. The plant quickly takes root if there are suitable conditions in the dune or peat zones, but it is also quickly overgrown: it is considered a 'pioneer species'. Because the corridor pipelines regularly need to be opened up, the overgrowth here has no chance to take hold, and the rare fen orchid can grow unfettered.

The natterjack toad is another pioneer species, also a rare one. These animals migrate to open sands such as vacant construction sites, and can then cause considerable difficulties for the construction work. Because of their protected status, they cannot simply be killed. But the pipeline corridors offer a solution: toad pools have been built in the residual spaces next to the actual pipeline corridor, between the grass and orchids, and marked by a number of boulders. These pools are shallow, concrete shells buried in the sand. Toads that find themselves on a site where construction is to take place are moved to these pools, in buckets. They enjoy the pools, according to the annual monitoring report. In these pipeline corridors, it seems that they have been able to form a stable population of around 400 adult animals.

– PPW –

The number of orchid species per square kilometre in
the city of Rotterdam
1
2
3
4
5
6

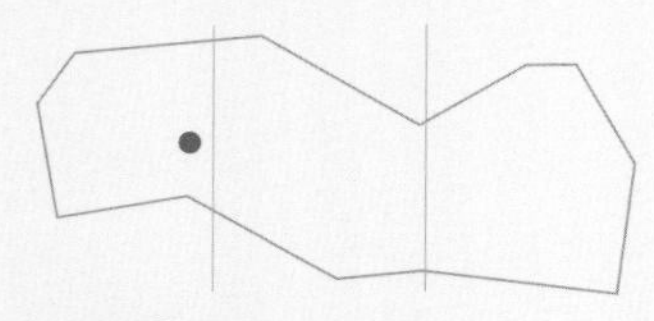

– 10 –

D'Arcyweg 76 \ Europoort Rotterdam

HARTELKANAAL

A Shield of Sand

Office buildings do not really belong in areas such as Europoort and the Maasvlakte. Almost all of the companies have a small office on their site, often built into the premises, but free-standing office buildings are rare. The old harbours near the city are much more suitable for the port-related companies that provide their services from behind a desk.

Nevertheless, a few of those solitary office buildings do exist here. The customs department, for example, needs such a building, as does the dredging depot the Slufter. The office of the BP Raffinaderij is another. It was built after an explosion at a refinery in Texas killed 15 workers, and oil companies took additional security measures.

In the port area, a 'security contour' regulation is in place. Outside of this area, any 'place-related risk' of greater than 10-6 is not allowed. That means that if a fictitious unprotected person were to spend a year somewhere outside of this contour, the risk that he or she would die as the result of a calamity is not greater than one in a million.

The urban and village areas lie outside of this safety contour, but the BP office is right in the middle of it. And there is no other way; the staff regularly has to commute back and forth between the office and the refinery.

That is why the design, made by architecture firm Group A, includes a number of special security measures. The office is hidden in an artificial dune, which has been planted with vegetation that refers to the dunes that grew here long ago. If there is an explosion, the thick layer of sand will absorb a large part of the shock, and the building's round shape ensures that the pressure wave will roll across the building. Extra-strong concrete was used in the construction. The office is divided into two compartments, with an atrium between them. Daylight enters the building via a roof made of anti-explosion glass. The front side the building has an open appearance, facing the green space of the Brielse Meer and the Hartelkanaal, but it too has been built using anti-explosion glass.

– PPW –

Refinery seen from the south (on the left in the aerial photo).

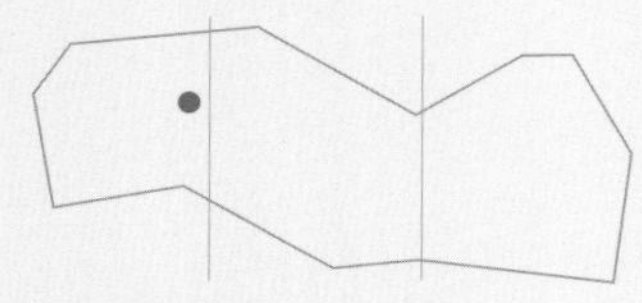

– 11 –

Markweg 131 \ Europoort Rotterdam

DINTELHAVEN

Terminal for Ore Giants

The name of the company Ertsoverslagbedrijf Europoort CV, better known as EECV, says it all: this is a facility for the transhipment of iron ore (about 20 million tonnes in 2013-2014) and coal (about 6 million tonnes). The story behind this terminal makes you realize that Rotterdam is largely a 'German' port.

Thanks to the presence of coal and iron ore in the ground along the Ruhr River, a tributary of the Rhine, this region in the west of Germany has developed, since the end of the nineteenth century, into one of the largest industrial centres in Europe. The steel industry in particular is booming in the Ruhr Region. For the supply of raw materials and the delivery of products, more and more use is being made of the Port of Rotterdam, which is located at the mouth of the Rhine. Especially with the newly dug New Waterway, Rotterdam offers a perfect connection to that other major industrial power, England.

Two world wars and one economic miracle later, mining in the Ruhr Region is waning, but the steel industry is still going strong, not least thanks to the steel used by Volkswagen, Opel, Mercedes and other German car brands. More and more, the raw material of iron ore needs to be brought in from elsewhere in the world. And the logical supply port for that purpose is Rotterdam.

Rotterdam has responded to this development by building a new ore terminal

in the Europoort area, and by digging the Eurogeul, which will enable the largest of the bulk ships to enter the port. This is necessary, because those ships are becoming big, bigger and biggest. In 2015, the *Berge Stahl*, which for years was the world's largest bulk carrier, now sails exclusively between the Brazilian port of Ponta da Madeira and Rotterdam. Nine or ten times a year, this ship delivers 360,000 tonnes of iron ore from the mines of the Vale company, located in the Amazon; it then sails back to Brazil, empty, for a new load. Barges cruise along the Rhine to ship the iron ore to the blast furnaces in the Ruhr Region, especially in Duisburg.

A few years ago, mining giant Vale had ships built that are even larger than the *Berge Stahl*. One of the ships in this series, the *Vale Rio de Janeiro*, moored at EECV in 2012.

The suggestion that Rotterdam is a 'German' port can in this case be interpreted quite literally: both EECV and the inland shipping company for the hinterland transport are owned by steel company ThyssenKrupp.

– FdK –

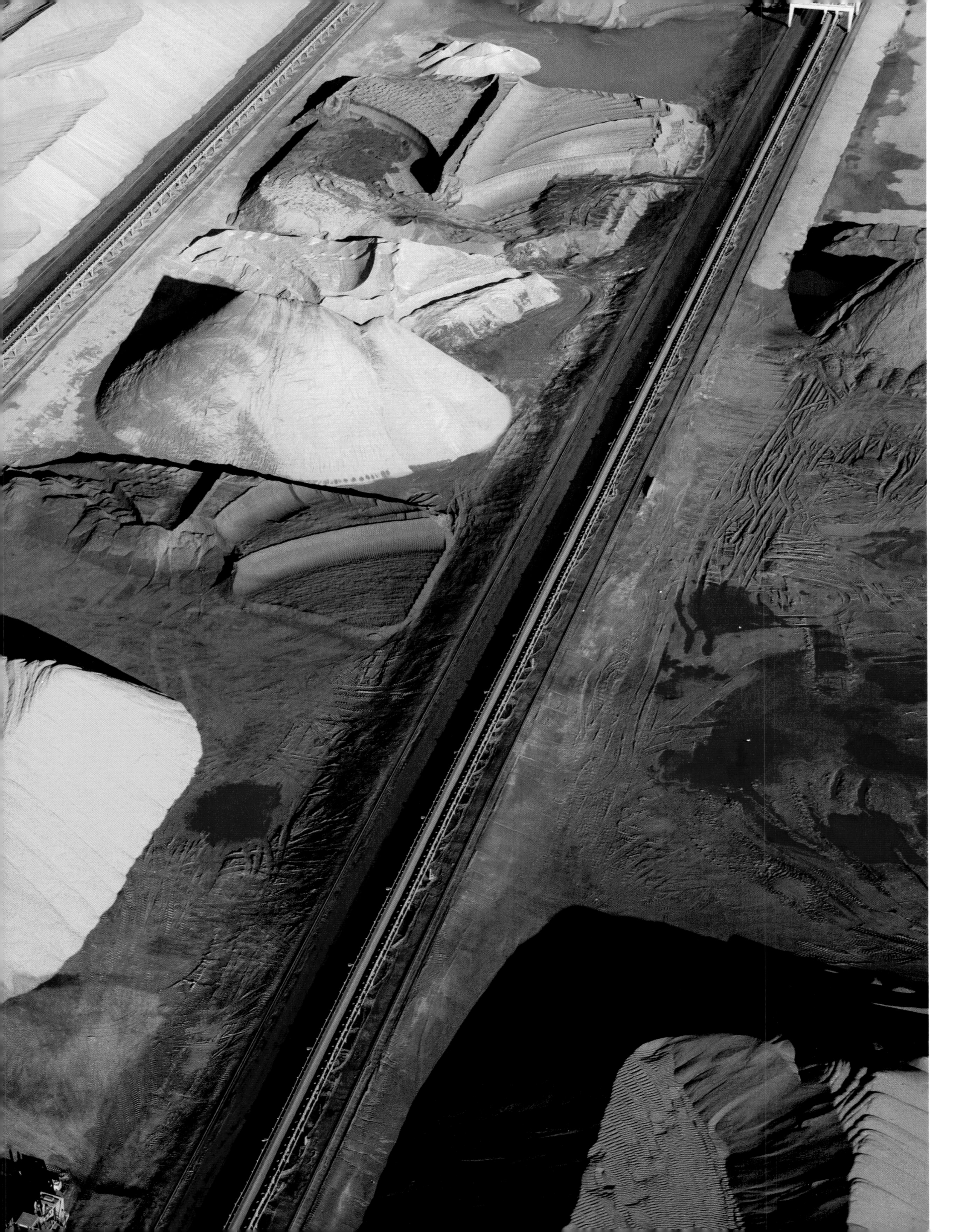

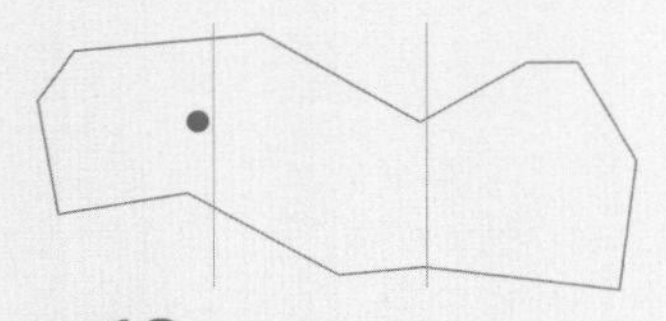

– 12 –

Elbeweg 117 \ Europoort Rotterdam

BENELUXHAVEN

A Hub for Dry Bulk

Trains run to and from the terminal of European Bulk Services (EBS) in Europoort to load and unload grain: grain that's used to bake your bread, in addition to raw materials to build a roof over your head and coal to heat your house. Ports play an important role in supplying people's daily needs. And that role is not a local one, but an international one.

The things that people need every day are transported in large quantities: in bulk. Take grain, for example: different types such as wheat, rye, barley, rice and corn travel the whole world over. These streams fluctuate greatly, because a crop failure in one part of the world can mean a shortage in another part, leading to extra demand in yet another part. Ships transport all those tonnes on the open sea, while inland this transport usually takes place via rail and inland waterways. The transhipment in ports, for example here in Rotterdam at EBS, is an industry in itself. The introduction of the elevator – a kind of vacuum cleaner for grain – in the early twentieth century was a major step forward in the expansion of scale.

Wherever there is transhipment, there is also storage. Every port has familiar landmarks in the form of silos, where the grain is transported in and out. But other types of dry bulk are also very recognizable: coal is piled high at another EBS terminal. This is also the case at EMO, which is a sister company of EBS as well as the largest coal terminal in Europe. Both companies are part of HES Beheer, which emerged from the Graan Elevator Maatschappij (GEM), and which is now a holding company that owns several dry bulk terminals, also beyond Rotterdam.

Agribulk can be divided into *food* for human consumption and *feed* for animals. A third category is becoming increasingly important for the Port of Rotterdam: *fuel*. A growing proportion of agricultural production serves as raw material for energy. Rapeseed and soybeans for biodiesel, for example, and grains for bioethanol. Rotterdam also ships more and more wood pellets, which is a form of wood waste that disappears into the power plants along the Maasvlakte.

– FdK –

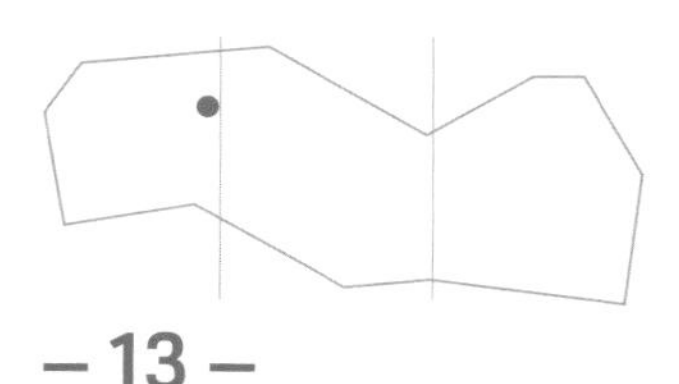

– 13 –

Noordzeeweg 1090 \ Rozenburg

KOP VAN DE LANDTONG

A Sunday Outing to the Port

When the harbours of the Europoort were built at the end of the 1950s, there was a need for a separate way of accessing the sea; the New Waterway was not deep enough. Excavation was not desired, because the Maas carries a lot of silt, and it would take too much effort to maintain the navigable channel at the proper depth. That is why the choice was made to dig the agrarian island of Rozenburg. Since then, an elongated peninsula has separated the two routes, like the median strip along a highway exit, with the New Waterway as the access to the Botlek and the old city ports, and the Caland Channel as the 'Europoort exit'.

Driving to and from this peninsula used to be a popular activity. It was 8 km ofstraight road, with the reward of reaching a wonderful place from which to spot ships: the round head of the peninsula. Beyond it was a green hill that contained fly ash, a waste product created after the burning of coal; here it was safely stored under foil and a thick layer of soil. Ship spotters frequently visited this peninsula, as did fishermen: the estuaries on both sides, the strong current, the great depth and the groynes that stuck out of the water all guarantee a good and varied catch. But it was certainly an inhospitable place.

The growth spurt that port experienced in the 1990s, including the plans for the Maasvlakte 2, was the reason to redevelop many areas in and around the port. It was a planning trade-off: the port could grow, but at the same time the residential and living environment in the region had to improve. This is referred to as 'dual purpose'. It led to a long series of measures to enhance the area's liveability, ranging from the conversion of 750 hectares of farmland into nature and recreation areas, to quieter asphalt and the renovation of a swimming pool.

It was at this point that the opportunities of the Landtong Rozenburg as a recreational and nature area came into view. Cycling and skating paths have been created, along with picnic spots along the route. Obstructing vegetation is removed and replaced by new plants. In the village of Rozenburg is a park designed with amenities such as a dog school, a mountain bike trail and an observation tower. Ramblers can easily encounter Scottish Highlander cattle and Konik horses here. And it's no longer necessary to travel 8 km there and back: the fast ferry now connects the head of this peninsula to the Hook of Holland.

The peninsula has become a great destination for a Sunday outing in the port. But ironically enough, it doesn't attract nearly as many people as the Maasvlakte 2 does, even though the recreational facilities here were intended to compensate for the construction of the Maasvlakte 2.

– PPW –

Line

DFDS SEAWAYS

Siebe Swart

AIR, LAND AND WATER

The Maasvlakte 2, Princess Alexiahaven and the Maasvlakte Beach.
A manmade beach and dunes as a 'soft seawall'.

Europoort, with the Maasvlakte on the horizon.
Dintelhaven Bridges of the A15, and the Betuweroute via the Hartelkanaal.

Maasvlakte with container terminals.
Europahaven with APM Terminals and the ECT along the Amazonehaven (on the right).

MAERSK
UBC

Maasvlakte, ECT Delta Terminal.
Fully-automated container cranes and unmanned vehicles.

826
ect
276
NELCON

Botlek, Brittanniëhaven.
Auto terminal of Cobelfret (C.RO Ports) with new cars.

Europoort, 7e Petroleumhaven.
Oil tankers moored at the Vopak Terminal.

Vopak

Rotterdam-Zuid, Katendrecht (De Kaap).
Maashaven and Rijnhaven (on the right).

Maasvlakte, Europaweg.
Freight train on the tracks of the Betuweroute, heading towards the Maasvlakte emplacement.

ESTIMATED TIME OF ARRIVAL

Port Development before 1940

Marinke Steenhuis

01

Romeyn de Hooghe and Johannes de Vou, map of Rotterdam (1694). The Waterstad – the area comprising of Wijnhaven, Leuvehaven and Scheepmakershaven – was the Maasvlakte of the seventeenth century. After the fall of Antwerp, in 1585 the city council had the sandbars in the river embanked. This marked the beginning of the expansion of the Port of Rotterdam.

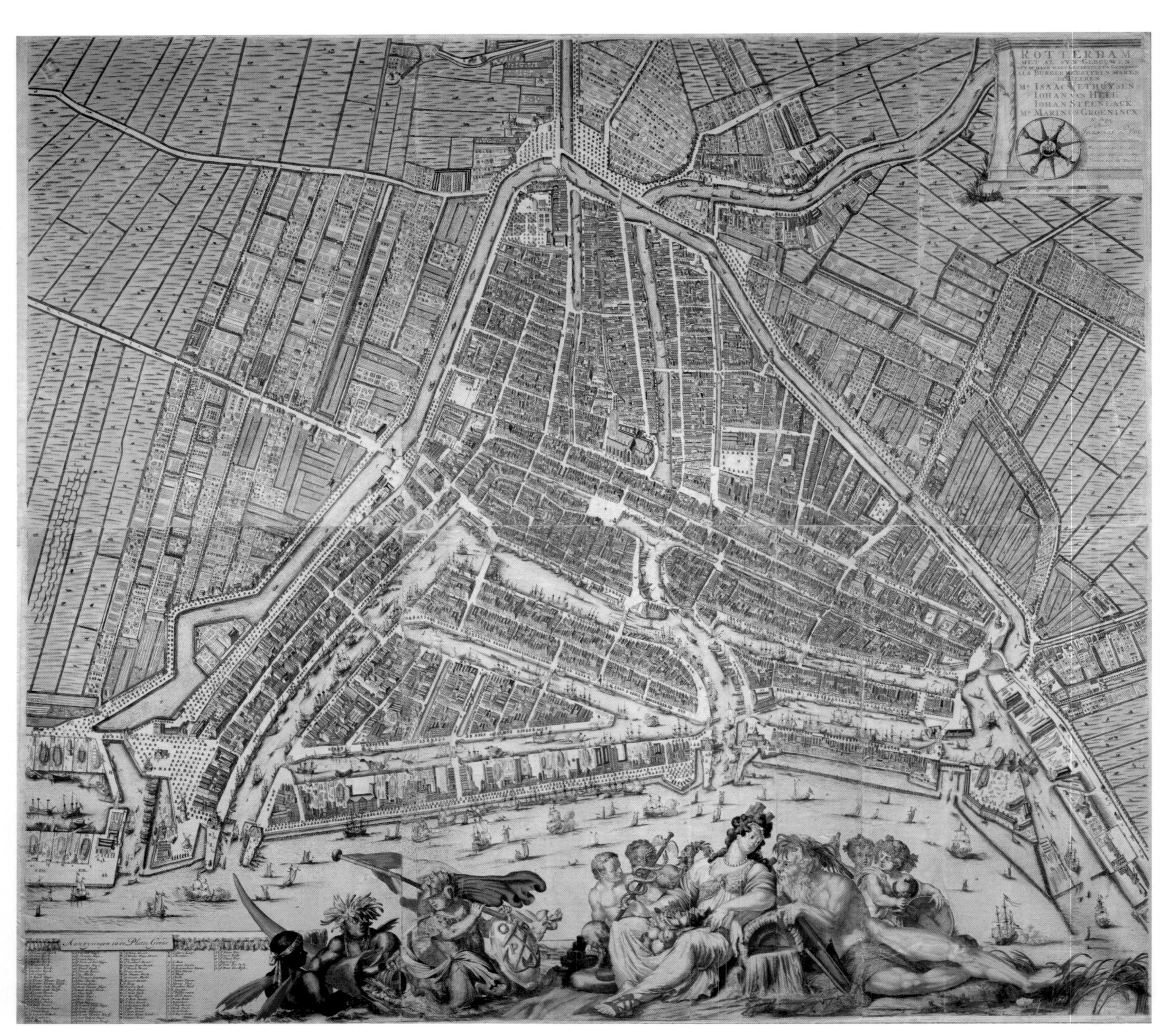

01

FOUR CENTURIES OF LAND RECLAMATION ON THE MAAS

Although the name Rotterdam refers to the damming of the Rotte River, the Rotte is just a tributary of the Rhine. It is that mighty river which here, in the estuary region of the Netherlands (known in this country as the *Delta*), reaches the sea, its two branches at this point being known as the Oude Maas and the Nieuwe Maas. This makes the Netherlands a unique meeting-place of the European hinterland with the oceans of the world. Since around 1850, the landscape of the Rhine Estuary (*Rijnmond*) has been erased and reconstructed time and time again, in sync with the waves of technological progress.
We are amazed in our day at the technical prowess of the Maasvlakte 2, but the original feat was the construction in the early seventeenth century of the harbours of this city, which being situated on a tidal river had to be built outside the protection of the dikes. With the same speed and courage as the recent construction of the Maasvlakte 2, which was put up in three years after 18 years of deliberations, the city councillors of Rotterdam seized their economic chance after the fall of Antwerp to the Habsburgs in 1585 and put up dikes around the sandbanks in the river's tidal range. This was the irrevocable beginning of the expansion of Rotterdam as a port city. For all the centuries since then, the businesses situated along the banks of the Maas and the New Waterway have been crucibles of innovation. The port landscape has been dug out, built on, renovated, demolished, built up again, shunted along, and finally made bare. That continual renewal has given rise to a supremely functional landscape that has, thanks to the next monumental work of the New Waterway of 1872, gradually crept seawards. The scale of the nineteenth-century port has now become the background scale of the Rotterdam skyline. Constructions that were colossal in their day are now barely noticeable. The grain elevators at the Maashaven, the imposing Koningshaven lifting bridge of 1927, the container terminal cranes on the Maasvlakte 2 that date from 2014 – all these are utilitarian icons that demonstrate how totally the continually changing port landscape has determined the culture and mentality of Rotterdam as a city.

CALAND'S BRIEFCASE

This was the briefcase in which hydraulic engineer Pieter Caland (1826-1902) carried his sketches for the spectacular New Waterway. Caland made a hydraulic design that caused the Maas to begin shifting its silted riverbed 30 km to the west of the city. The route was officially opened in 1872.

A GREAT LEAP FORWARD: THE PORT FROM 1850 TO 1940

We are standing on the Gatdam in Brielle, gazing outwards to sea. There is nothing here now to indicate that we are right in the middle of what used to be the approach to Rotterdam port. To the left is the pretty island of Oost-Voorne, home to so many port workers; to our right, the Slufter, a hermetically-sealed dump for contaminated silt from the port. De Pit, De Beer, De Zekken. De Spleet, Het Scheur, De Geul. De Binnenvlakte, de Droogte van de Maas. These names of stretches of water are not very meaningful even to locals any more, but for centuries they were the essence of day-to-day Rotterdam shipping. Access to the sea down the Nieuwe Maas was already being impeded around 1800 due to silting-up. Consequently, until the New Waterway was dug, ships had to make a detour and took days to reach the open sea. Captain Pieter van der Hoog (1835-1906), from Krimpen aan de Lek east of the city, set sail from the southern Dutch port of Brouwershaven in 1863 aboard his three-master, the *Bastiaan Pot*, bound for Australia: a voyage of nearly 140 days at sea. Yet even before he reached Brouwershaven, he had been aboard the vessel for quite

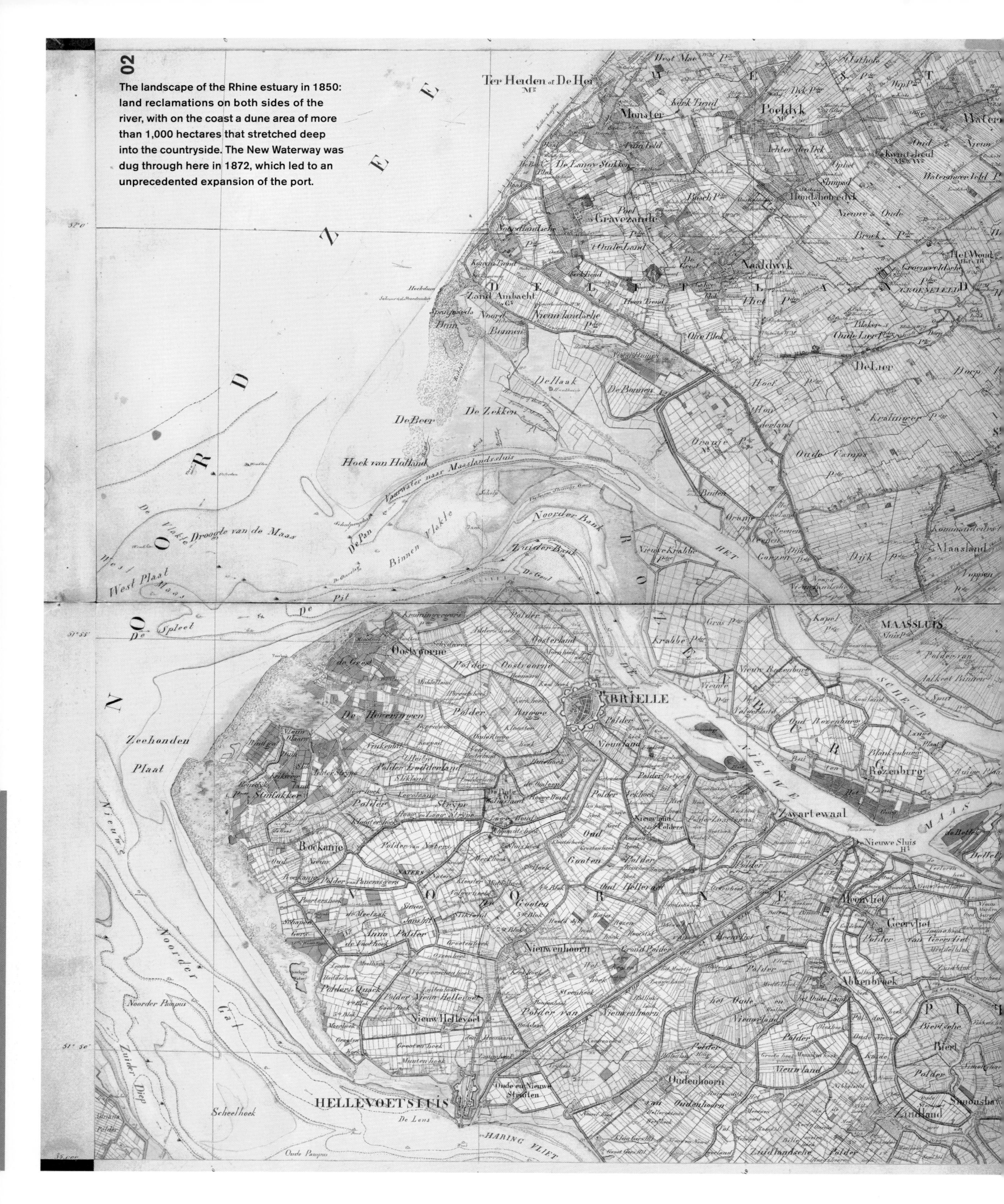

02

The landscape of the Rhine estuary in 1850: land reclamations on both sides of the river, with on the coast a dune area of more than 1,000 hectares that stretched deep into the countryside. The New Waterway was dug through here in 1872, which led to an unprecedented expansion of the port.

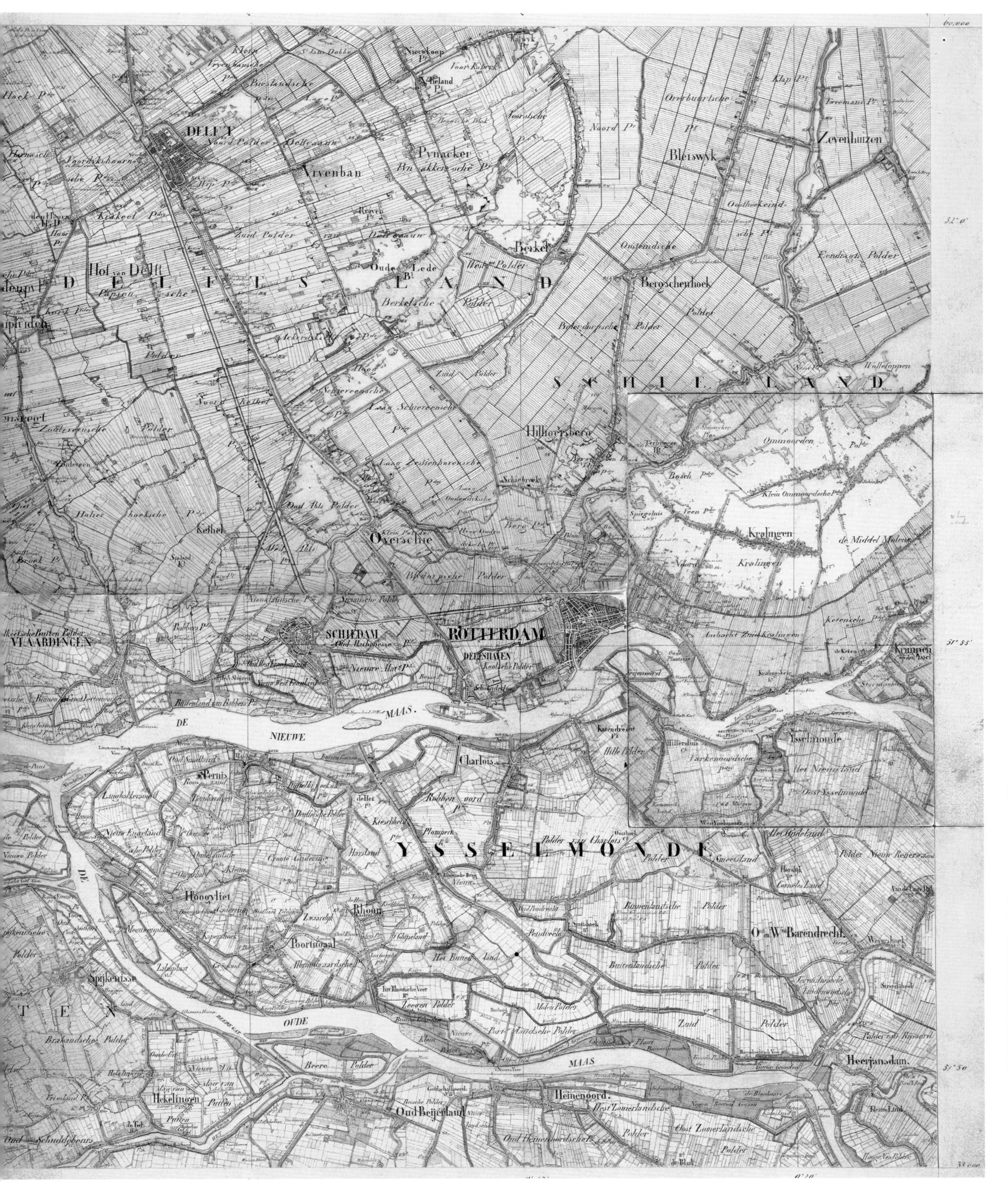
DELFT
Pynacker
Bleiswyk
Zevenhuizen
Hof van Delft
D E L F L A N D
Berkel
Bergschenhoek
S C H I E L A N D
Hillegersberg
Overschie
Kralingen
VLAARDINGEN
SCHIEDAM
ROTTERDAM
DELFSHAVEN
DE NIEUWE MAAS
Charlois
Ysselmonde
Pernis
Y S S E L M O N D E
Hoogvliet
Rhoon
Poortugaal
O. en W. Barendrecht
Spijkenisse
OUDE MAAS
Heerjansdam
Hekelingen
Oud Beijerland
Heinenoord

03 – 06

In the nineteenth century, the port and the city were still united. The silhouette of the city was shaped by the stately offices and homes of the merchants and ship owners, the water was teeming with ships, and the docks were busy with the coming and going of porters. The smell of tar and fish was never far away.

03

04

05

Mouth of the Rhine, Rotterdam, Holland.
Series 1894, by Bert Underwood.

06

some days. Setting out from Krimpen aan de Lek, he had to set a course between the polders and salmon fisheries of IJsselmonde, the steam-powered factory of Feijenoord and the majestic mercantile townhouses along the Boompjes, past the brand-new Maas Park and further down the Nieuwe Maas until he reached Rotterdam port proper. Signal boats had to be specially hired to ensure he did not run aground on a sandbank. Reaching Vlaardingen to the west of the city, he found his passage blocked: the waters of Het Scheur had ebbed away to a bank on which he would be stranded. The captain was thus obliged to veer off to the south, around the tip of Rozenburg Island. Horses towed his ship down the Voorne Canal. Arriving in Hellevoetsluis on the other side of Voorne, he had to sail down the Haringvliet and around the tip of Goeree to reach Zeeland and the departure port of Brouwershaven. There, Van der Hoog finally had clear passage to the sea, but he still had to wait for favourable winds to blow in the right direction to set sail.

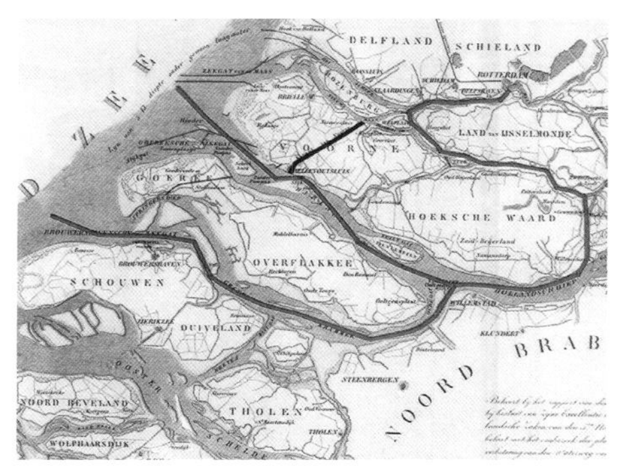

NEW WATERWAY: PUTTING THE RIVER TO USE

Ten years later, attempting the same journey that Van der Hoog had made was a completely different experience. Rijkswaterstaat, which manages Dutch waterways, decided to dig out Het Scheur into a 4.3-km-long straight channel to the sea, following a design by Engineer Pieter Caland. The brilliance of Caland's design was that he let the river take the strain: his engineering plan involved the Maas beginning to carry its own sandy bed out to sea, 30 km west of the city. No sea sluices were necessary, and the load line of the river would keep the waterway at the proper depth. It was in 1872 that the first ships set out to sea down this route. Steamships, which were coming into their own in this era, made the New Waterway a good investment by using the route. Besides the old and slightly stuck-in-their-ways merchant class of Rotterdam, there now arose a new class of businessmen in the city that was eager to invest in the port, especially in the flexibility of transhipment trade. The government gave Amsterdam the North Sea Canal as a consolation prize, but Rotterdam's new improved connection with Germany down the Rhine was what really set the pace. Rotterdam had become the gateway to the world, with Hook of Holland as the first and last settlement along the route.

SALMON FISHING INDUSTRY

As early as the fourteenth century, the salmon fishing industry was based in IJsselmonde. Because salt and fresh water mixed together at this location in the Nieuwe Maas, it was an area abundant with salmon. The fishing took place using large nets in the rivers, much like it is done today in the North Sea. Horses or steam engines would pull a large net (called a *zegen*, or blessing) though the river, from the riverbank. Industrial pollution gradually put an end to salmon fishing here around 1930.

More than any other intervention, it was the digging of the New Waterway that inaugurated Rotterdam's modern port era. It was here that the city's rhythm as a centre of transfers arose, thanks to the deep navigable waters and the continuous flow of goods from and to sea and the hinterland. Yet the New Waterway would not have had any significance at all had it not been for the trade treaties signed in that age, which were greatly modernizing international business. The Revised Convention for Navigation of the Rhine was signed at Mannheim in 1868. This treaty guaranteed trade ships unhampered passage up and down the mighty river. In this era of land wars and the quest for a balance between the European great powers, the countries surrounding the Rhine well understood that a free, common transportation market was essential for European prosperity to continue growing. This treaty, nearly a century older than the origins of the European Union, gave Europe an enormous competitive edge. A comparable agreement in our time would be the Port Base system, under which vessels no longer have to wait for endless rubber

07

Before the arrival of the New Waterway, ships had to sail around to reach the North Sea. The route not only took a lot of time, but once ships arrived in the Brouwershaven in Zeeland, the skipper still had to wait for the right tidal and wind conditions before continuing his journey.

08

Upon the completion of the New Waterway, routes that often required detours of hundreds of kilometres were replaced by a short, deep route of only 30 km that also allowed access to a new invention: the steamship.

09

The New Waterway turned the port into a modern transhipment port, whereby the free trade treaties created a common transport market.

10

The New Waterway, an excavation through 4.3 km of dunes that made Rotterdam the biggest port in the world.

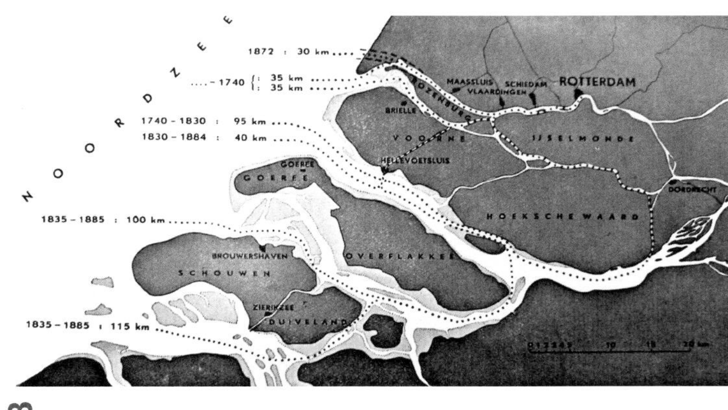

08

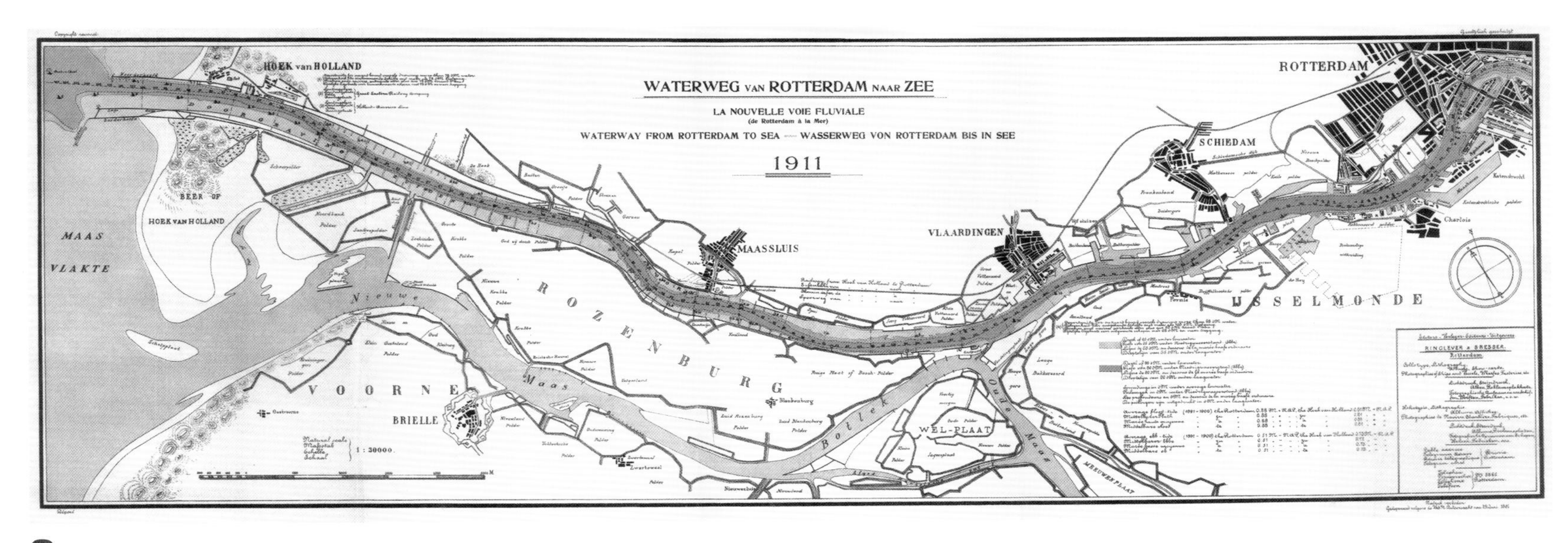

09

10

stamps before loading or unloading, but instead are given electronic access to the counters at the Port of Rotterdam to process customs declarations and cargo manifests using a single shared standard. In transhipment, the essence of everything in the port is speed and efficiency.

'AN IMMENSE FLEET HELD CAPTIVE IN A CITY'

The port was in need of storage space. Around the mid-point of the nineteenth century, W.N. Rose, Master of City Works, had made a start on this by integrating the spatial needs of the port and of the residential city into a single model in which the architectural profiles of the housing and port areas were attuned to each other and in which work was undertaken to modernize the water management and sewage system. After 1850, Rotterdam began to boast upper-class canal streets (*singels*) and a new neighbourhood, Het Nieuwe Werck, was built with a park lying along the Maas (although the city council had initially intended for that land to host an abattoir). Rotterdam entrepreneurs were among the very first in the world to concern themselves with public works in their city. This was doubtless due to the city's reputation for weak political leadership. The huge increase in demand for berths prompted the Veerhaven and Westerhaven to be dug at the initiative of the city's Chamber of Commerce, which had been founded in 1803.

Although the shortage of berths and the shortfall in supply of new factory premises was enormous, it was not until 1870 that the city first encroached onto the south bank of the Maas. It took the construction by the national government of a swinging railway bridge high above the river, designed not to impede shipping, for the locals to pluck up the courage to do likewise and cross over to the south side, once feared as the home of the gallows and the plague house. The first high railway bridge was followed by a rotating railway bridge above the newly-dug Koningshaven; this delineated a spit of land which became the Noordereiland. The city council lacked the resources, however, to build harbour basins in Feijenoord, since the councillors did not dare to raise local taxes to fund the project in advance.

GUTS AND ENTREPRENEURIAL SPIRIT

The rise of Rotterdam as a transhipment port owes everything to its entrepreneurs, men who had the guts to invest in what were sometimes immense projects. After the New Waterway had been dug, all kinds of new arrivals washed up in the city, earning a crust at some point or other in the chain of the international goods trade. They were commission merchants, hauliers, shipbrokers and steamship owners, for instance. It was in Feijenoord in 1871 that the adventurer and pioneer Lodewijk Pincoffs (1827-1911) founded the Rotterdamsche Handelsvereeniging (RHV), the city's trade association, whose purpose was to arrange loading and unloading and storage for ships' cargoes in premises in the newly-laid harbours around Feijenoord. Pincoffs anticipated that he would be able to meet a need arising from the lack of berths at that time, a lack that had been forcing seagoing ships to resort to the time-consuming and expensive measure of unloading their cargoes on the river into shore barges. He drummed up millions of guilders of investment capital through a share issue to Rotterdam families and other investors. In December 1872, the city council decided that a parcel of land in Feijenoord would be partly

IMMENSE FLEET, CAPTURED IN A CITY

When Italian traveller Edmondo de Amicis visited Rotterdam's Waterstad in 1874, he wrote: 'On every side, between the houses, above the rooftops, between the trees, you can see masts, flags, sails, and rigging; this city is a seaport, an immense fleet, captured in a city.'

MARTEN MEES

Together with Lodewijk Pincoffs, Marten Mees was part of a core group of businessmen in Rotterdam. He was involved in the establishment of the Rotterdamsche Bank, which eventually merged into ABN AMRO. Until his death in 1917, he was the chairman of the supervisory board of the Holland America Line.

11

A city that combines living and working: the Eerste Gemeentelijke HBS on the Kortenaerstraat, various hospitals, the Museum Boymans at the head of the Schiedamsedijk, a theatre and bank buildings on the Schiedamsedijk, and a post office at the bottom of the dike along the Vasteland. The Scottish church with a workhouse on either side of the Vasteland (rebuilt after May 1940 on the Schiedamsevest).

12

This postcard of the Oudehoofdplein shows colliding worlds and eras. Goods are loaded and unloaded on the Oosterkade, and right next door a long narrow park has been built. Here both steamships and sailing ships are shown making their way on the Maas.

13

In 1871, the former port estuary of Rotterdam, the Binnenrotte, was filled in to make way for the construction of the railway viaduct along the Rotterdam-Dordrecht line.

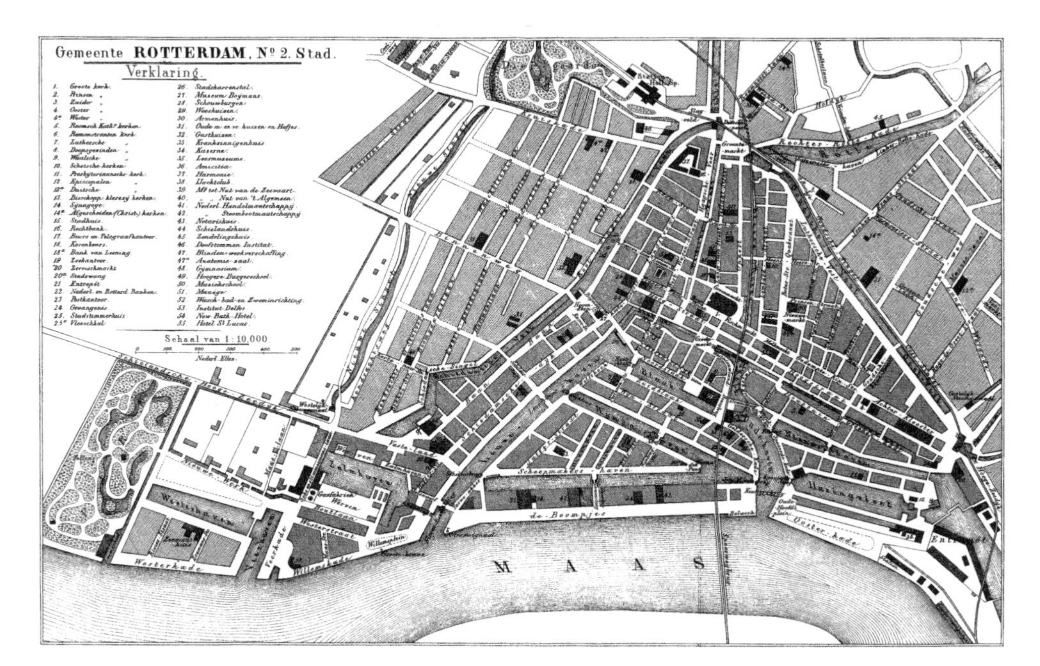

11

12

13

14

15

The gatehouse was the headquarters of the Rotterdamsche Handelsvereeniging (Rotterdam Trade Association), which was founded in 1872 by Lodewijk Pincoffs.

15

As the largest employer in the Port of Rotterdam, Daniël George van Beuningen was of great significance to the city. He lived in the Scheepvaartkwartier (Maritime Quarter) until the Second World War, and his large schooner *Vigilanter* was docked in the Veerhaven right in front of his office.

tendered out on an extended lease and partly sold to the Handelsvereeniging. Just a couple of years later, three planned new inner harbours were a reality: the Binnenhaven, the Entrepothaven and, at the railhead, the Spoorweghaven. In May 1879, Pincoffs suddenly fled to the United States, leaving the Handelsvereeniging saddled with debts and becoming the talk of the enraged city. A huge fraud came to light and the result was bankruptcy. Although 17 Rotterdam families lost a total of 14 million guilders, this debacle proved to be the catalyst for a new era. The City Council, which had always been so cautious, now took the initiative under the influential shipping line owner and councillor Jan Hudig (1838-1924) to take over the bankrupt estate for 4 million guilders. Thus the city now had its Gemeentelijke Handelsinrichtingen (Municipal Commercial Establishment). From 1882 onwards, then, Rotterdam City Council was in business for quay hire, quickly augmented by rentals of cranes and other materials.

The markets and offices of shipping lines, captains of industry, insurers and trading houses were across the river from Feijenoord, in a part of the old city that dates from after 1900. The Wijnhaven was the address for notaries and assurance companies; Boompjes was the neighbourhood of merchants and shipping line owners. The Veerhaven played host to the office of the Steenkolen Handels-Vereeniging coal traders' association, and moored by their front door was the *Vigilanter*, a schooner owned by their director, D.G. van Beuningen. The story is that Van Beuningen had decided to park his grand ship there after having been refused membership of De Maas, the Royal Rowing and Sailing Association, whose premises stood just a few dozen yards away, directly facing his association. In the post-1900 city, every working day had a set rhythm, one quite different from ours. Office workers clocked in early, as did the dockers, and at half past one in the afternoon the streets were thick with traders making their way up to the Stock Exchange. The offices closed at half past three, and then it was time for the gentlemen to take their wives out socializing in the park beside the Maas. The waterways in the city were still tidal, rising and sinking twice each day, and when the wind was blowing from the north-west or there was a flood tide, the main squares of Blaak, Beursplein and Grote Markt had water lapping through them. The view over the Maas from Boompjes went on forever, and fields of rapeseed blossomed on the Noordereiland. It was not long before the rhythms of Rotterdam city life were to change drastically.

SPECIALIZATION AND EFFICIENCY: A HARBOUR FOR EVERY PURPOSE

It was the Head of Municipal Works, G.J. de Jongh (1845-1917), who brought Rotterdam into the new era. De Jongh, who had been trained as an army engineer, re-envisaged the port and the city as two worlds that met at one spectacular point. A larger-than-life character, De Jongh persuaded councillors at lavish dinners of the need for another and yet another new harbour basin. De Jongh did something else besides: he concentrated each raw material in a single harbour, which obliged companies to compete on efficiency and innovation. Even during its construction in 1887, the Rijnhaven changed its function from the original plan. Instead of being a harbour to keep inland waterway vessels in during ice drifts, it ended up being a transhipment harbour for what was in those days a veritable spectacle to behold: direct shipboard-to-shipboard handling. Katendrecht, an ancient village of 3,500

BALCONY ON THE MAAS

With land from the newly dug Tweede Katendrechtsehaven, the 'Bult De Jongh' or Parkheuvel (park hill) was built along the river in 1896. It was named after the director of Public Works, G.J. de Jongh. From this balcony on the Maas, Rotterdammers enjoyed a spectacular view of the harbour dynamics.

16

Loft apartments are built atop the Fenixloodsen in the Rijnhaven. Cultural institutions are now located in the old hangars.

inhabitants, had to make way from 1895 onwards for the Maashaven, where De Jongh was determined to concentrate all the grain transhipment firms. Threading their way through the Fenixloodsen (Phoenix Warehouses), which in our day are being repurposed as large residential complexes, were two cargo railway lines. The transhipment of mixed cargoes was concentrated on the right bank of the Maas, in the Parkhaven (built 1890), the St-Jobshaven (1906) and the Schiehaven (1909).

TRANSLATING WORDS INTO ACTIONS

De Jongh's new overarching concept made the Rotterdam of the dawn of the twentieth century a city of unprecedented vitality and explosive growth. With transhipment turnover booming, the city became awash with nouveau-riche port industrialists such as D.G. van Beuningen, the coal association head whom we encountered earlier. It was in this period that he amassed his famed art collection on the back of his coal wealth, which has since formed the core of Museum Boijmans Van Beuningen. The charge has been levelled at De Jongh that he did not give enough weight to the residential aspects of the city; that his eye for beauty and for urban harmony was largely confined to the port activities. True enough, everything about his approach seems to have been aimed at the greater glory of the port: for instance, he waxed lyrical about the vista afforded by a railway bridge, with a panoramic view of the bustling marshalling yards. It was with De Jongh that the Rotterdammers made *geen woorden maar daden* (put your money where your mouth is) their can-do city creed, turning their attention to a new industrial and industrious model of beauty. Rotterdam firms pulled out all the stops to speed up the loading and unloading of cargo. A walk along the Maas at any time between 1900 and 1940, at any time of the day or night, provided a breath-taking sense of theatre as the ballet of boats and industrial installations was framed by the enormous warehouses along the riverbanks. Wondrous antediluvian monsters of riveted steel bobbed up and down in the Rijnhaven. These were waterborne grain elevators and floating dry docks, kept there to reduce transhipment times to a minimum, and this in an age when telephony and telegraphy were enabling ever more rapid communication. In time, this dramatic port landscape put Rotterdam on the early twentieth-century tourist map. Fop Smit shipping line

THE WAGENVEER

The Wagenveer was a ferry across the Nieuwe Maas in Rotterdam. The ferry could be customized to the tides, with a deck that could be moved up or down by 1.5 m.

17

In 1883, two municipal docks were built. These hulking structures were both 30 m high; one was 90 m long, the other 45 m long. They were able to float on the water, which made repair and inspection of large ships much easier.

cleverly made a bob or two out of this interest by publishing a slender volume in 1913, *Wandelvaart door de Rotterdamsche havens* (A walking tour through the harbours of Rotterdam). The text takes the reader on board the vessels and documents the new self-consciousness and pride of this city, whose profile had transformed in the blink of an eye. As the 1913 guide puts it, the whole city appeared at that time to be 'a cauldron of simmering commercial energies that has reached the boil'. During the loading and unloading of ships, coal transporters with a 'Jacob's ladder' at their prow (a fixed tube mounted at a slant) cascaded coal into the seagoing ships' scuttles. The whole extent of what Rotterdammers now regard as downtown was still 100 per cent port space at this time: from the Lloyd Pier ships departed for the Dutch East Indies (now Indonesia); the Schiecentrale had just come into use; and there were 2,000 people working at the Wilton Wharf in Feijenoord. To take 1910 as an example, the Grain Elevator Company received delivery of 4,400 million kg of grain in its 16 immense elevators that year, with a total value of 300 guilders. The grain was stored in a silo entirely made of reinforced concrete, a facility which in the years around 2000 found itself hosting mega-parties under the banner *Now&Wow*. 20,000 tonnes of grain could be stored in the 42 octagonal cells of this facility.

So De Jongh's spatial and economic vision had been fleshed out in fewer than 15 boom years. The north bank was where the fixed transport lines and mixed cargo traffic was concentrated; the south bank specialized in bulk products, namely ores, grain, wood and coal, which were transhipped in the newly-dug harbours of Waalhaven, Maashaven and Rijnhaven. It is from this time that Rotterdam's modern DNA and the character of its riverbanks date.

In 1869, at the age of 30, Jan Hudig was elected to the city council with more than 80 per cent of the vote. He was an alderman from 1899 to 1909. This illustration shows (port) innovations such as the underwater quay, the floating dry dock and the telegraph, as well as newer facilities such as the hospital, the gasworks and schools.

19

From 1903 to 1921, the NV Wilton's Machinefabriek en Scheepswerf was located on the Westkousdijk, close to historical Delfshaven. In the background, on the Sluisjesdijk, is the refinery of the Koninklijke Nederlandsche Maatschappij tot Exploitatie van Petroleumbronnen in Nederlandsch-Indië (KNPM), a forerunner of Shell.

20

In 1912, the floating grain elevators in the Maashaven unloaded 90 per cent of the total grain supply. Like a giant vacuum cleaner, the (floating) grain elevator empties out the hold of a ship at a very high pace. The introduction of the grain elevators meant a very drastic change. The old, labour-intensive form of grain handling provided employment to many port workers, who became unemployed with the arrival of the grain elevator.

THE VERSATILITY OF DE JONGH'S HARBOURS

The colossal Waalhaven, which was dug in phrases from 1912 to 1922, is currently undergoing redevelopment to serve as a maritime services centre. The Waalhaven project is exemplary of how Rotterdam as a port has been staying abreast of the times, a process that continues unabated today. What happened in the Rotterdam of a century ago was due to the convergence of several circumstances: more room was needed because of the growth in shipping, but also because new legislation was shortening the working day, obliging ships to stay moored for longer. In addition, cranes were increasingly taking on the bulk of loading and unloading from manual labour, and these cranes were more stable and safer standing on a quayside than they would be floating in the river.

Some of the foremost captains of Rotterdam industry decided to base their businesses on the Waalhaven. The brothers Frans and Cornelis Swarttouw, who managed stevedores for the transhipment of ores and other bulk goods, but also D.G. van Beuningen, then the city's largest employer, whose coal association fuelled the ships. Frans Swarttouw had presciently written as early as 1887, and in English at that, that the essence of his business must be 'quick dispatch'. In 1902, he was Rotterdam's leading stevedore boss. After 1912, his premises on the Waalhaven were replete with four imposing pontoons for loading and unloading. For decades, transhipment using manpower was the core activity on the south bank. It is a testament to the big thinking of its designers that the Waalhaven was large-scale enough to allow for a seamless transition by the stevedore companies after the containerization of the industry in 1966. In 1920, the Waalhaven Aerodrome, in constant use

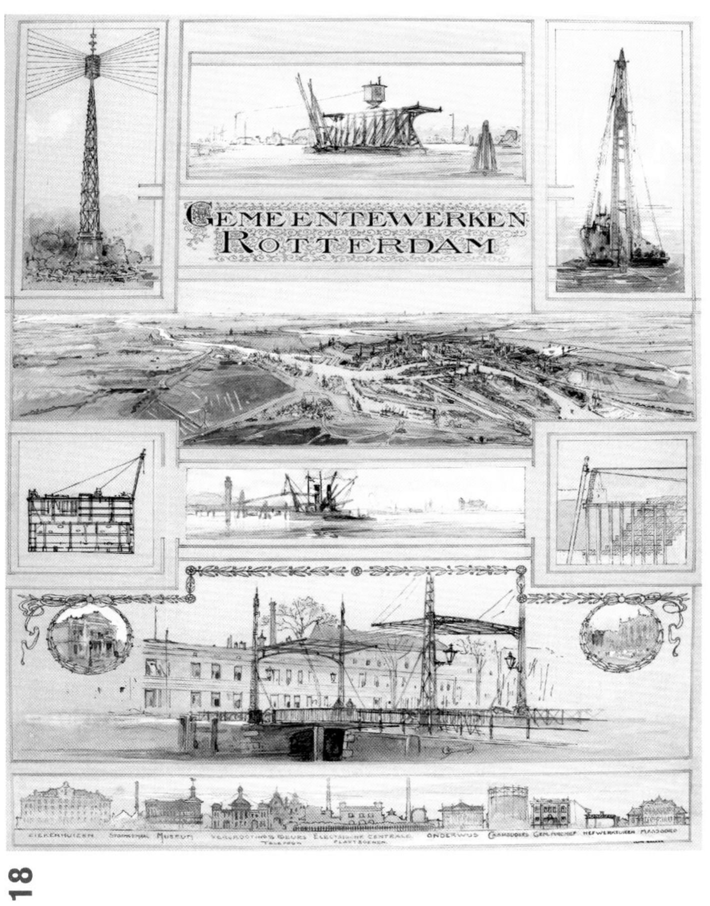

18

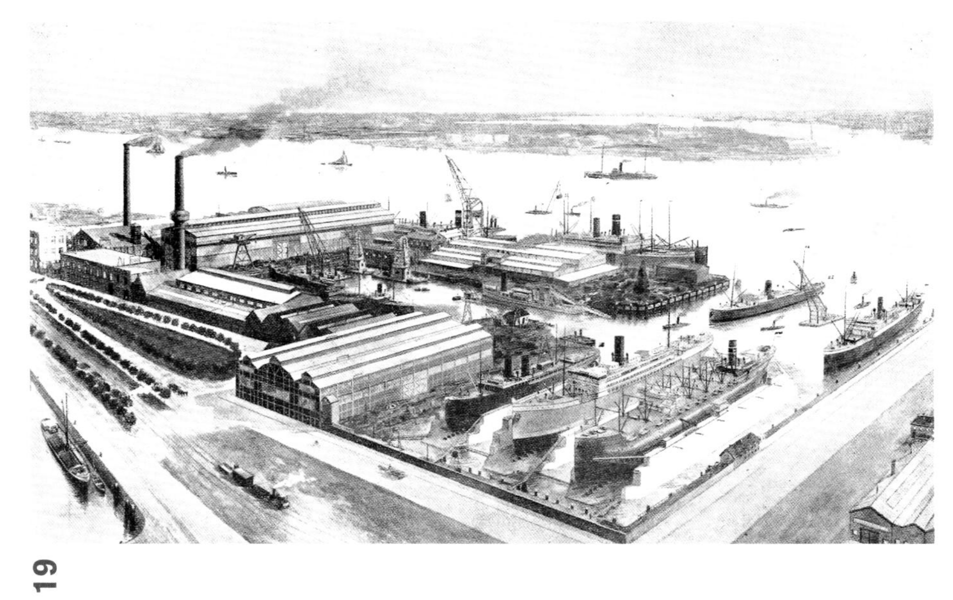

19

20

21

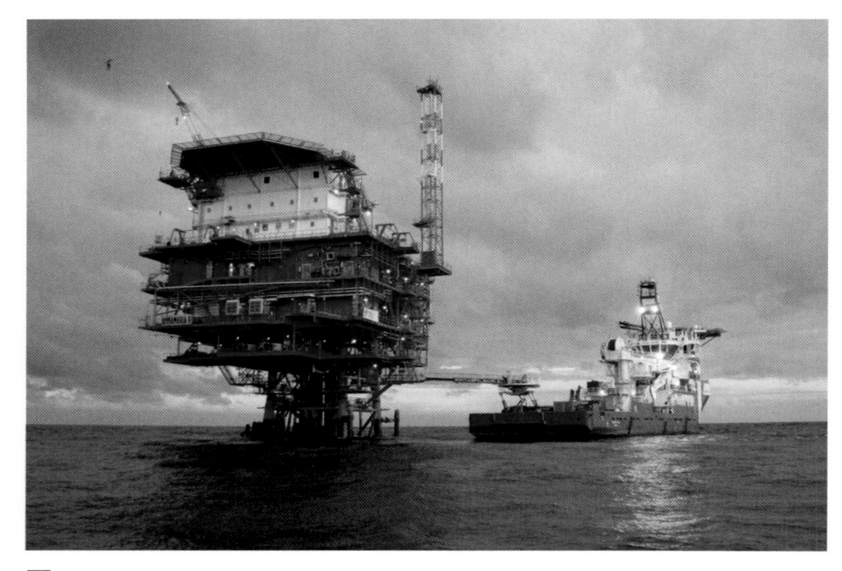

21

Several years ago, young engineers from Delft founded the company Ampelmann, which is located on the RDM site. The Ampelmann is a stabilizer at sea, which works like a flight simulator in reverse. It detects all the movements of a ship, and compensates for these by using six hydraulic cylinders. This ensures that the top side of the Ampelmann is completely motionless, and guarantees safe access to the sea.

ROTTERDAM

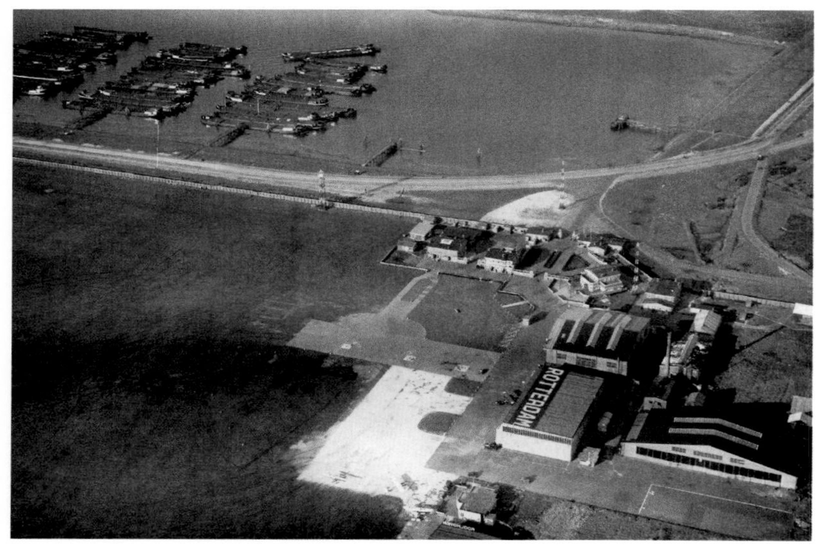

22

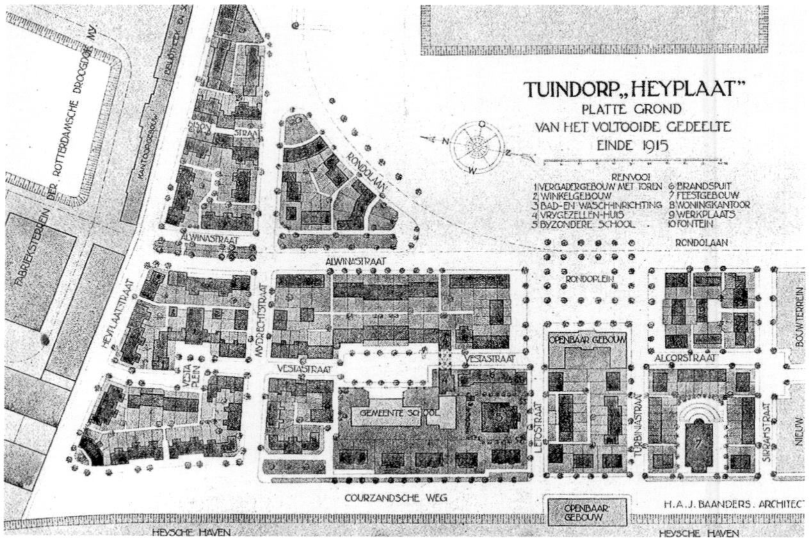

TUINDORP „HEYPLAAT"
PLATTE GROND
VAN HET VOLTOOIDE GEDEELTE
EINDE 1915
RENVOOI
1 VERGADERGEBOUW MET TOREN
2 WINKELGEBOUW
3 BAD-EN WASCHINRICHTING
4 VRYGEZELLEN-HUIS
5 BYZONDERE SCHOOL
6 BRANDSPUIT
7 FEESTGEBOUW
8 WONINGKANTOOR
9 WERKPLAATS
10 FONTEIN
FABRIEKSTERREIN DER ROTTERDAMSCHE DROOGDOK MY.
KANTOORGEBOUW
HEYPLAATSTRAAT
ALWINASTRAAT
RONDOLAAN
RONDOPLEIN
VESTASTRAAT
VESTA PLEIN
GEMEENTE SCHOOL
OPENBAAR GEBOUW
ALCORSTRAAT
LETOSTRAAT
TURBINIASTRAAT
SIRRAMSTRAAT
BOUWTERREIN
NIEUW
COURZANDSCHE WEG
H.A.J. BAANDERS. ARCHITECT
HEYSCHE HAVEN
HEYSCHE HAVEN

23

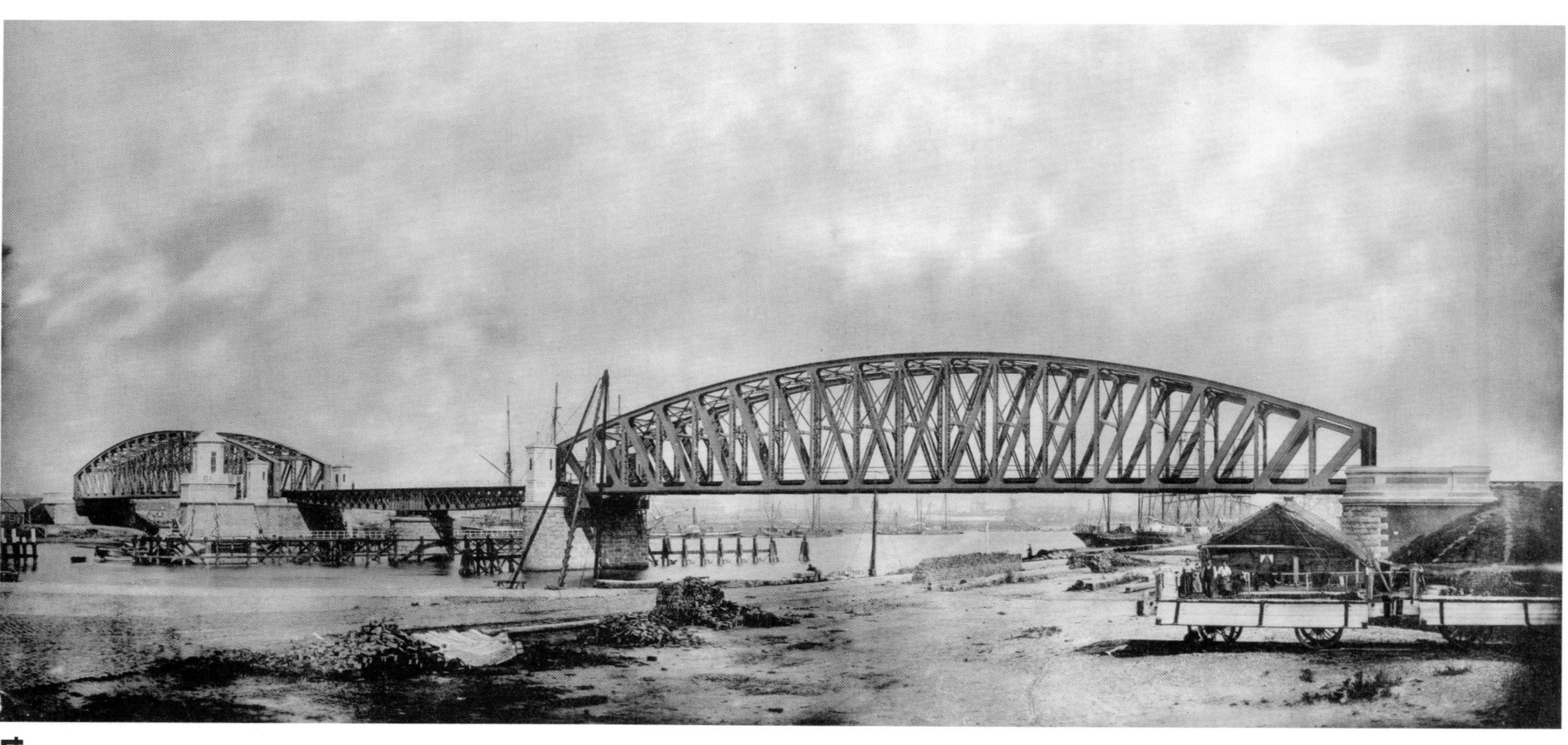

24

22

The Rotterdam airfield Waalhaven was opened on 26 July 1920. It was built on the first piece of property that was made available by the excavation of the Waalhaven. One of the directors of the Van Nellefabriek, M.A.G. van der Leeuw (1894-1936), flew daily from this airfield to his home in Rockanje for lunch. In 1936 he was killed in Africa when approaching an airport in thick fog.

Tuindorp Heijplaat (Heijplaat garden village) was designed by P. Verhagen, the leader of Stedenbouw Gemeente Rotterdam (Town Planning, City of Rotterdam) and H.A.J. Baanders (architect) as a residential area for the workers of the Rotterdamsche Droogdok Maatschappij. The workers lived next to their place of work. Today, the isolated location of Heijplaat, coupled with the RDM Creative Campus, exerts an attractive force, but back then it meant long travel distances to the centre, which resulted in a close-knit 'village in the city'.

A railway bridge over Rotterdam's Koningshaven in 1876, photographed by German photographer Johann Georg Hameter (1838-1885). By placing the rails at an elevated height, the shipping traffic below was not interrupted.

USING WORK TO CREATE WORK

In 1912, the dredged silt from the Waalhaven was transported through a system of pipes to the banks of the Kralingse Plas, to literally build a floor for the vegetation. Using work to create work seems to have been invented in Rotterdam. In 1920, the Rotterdamsche Kring (Rotterdam Circle), a cultural network of influential Rotterdam, asked the firm Granpré Molière, Verhagen en Kok to come up with a design for a recreational area. It was the first park design that did not attempt to create a romantic park construction, but instead made the existing polder landscape suitable for large groups of urbanites. In the city of the interwar period, social and physical engineering went hand in hand.

by the top class of port figures, was situated where the Waalhaven-Zuid industrial estate now stands. Sandwiched in between Eemhaven and Waalhaven is the garden village of Heijplaat, built by the directors of the Rotterdam Dry Dock Company (RDM) for their workers in 1914. A century on, it is still a glorious oasis of calm amid the industrial clamour, and nowadays is also the base of operations for creative entrepreneurs originating from the RDM Campus, a location where education is fused with innovation firms such as Ampelmann and the robotics lab Hyperbody. So it is that each stage of modernization has succeeded the last in this same spot.

PORT CAPITALISTS AND SHIPPING BARONS

Many of Rotterdam's residential neighbourhoods, too, have their history bound up with the growth of the port. Crooswijk (built 1913), Spangen (1914), Beukelsdijk (1914) and the later parts of Blijdorp (1914) postdate De Jongh's retirement in 1910 and were designed by a newly-recruited urban planner, Pieter Verhagen (1882-1950). However, besides him, the new class of merchants in the port were increasingly shouldering their responsibilities to house the thousands of working men who had flocked to Rotterdam to seek employment in their businesses. This was partly out of enlightened self-interest, since (as the thinking went) a man who would sometimes have to wait around to see whether work was on offer, who did not have a decent home, who was offered no cultural stimuli or whose children were not being schooled properly would slip into drunkenness or develop the habit of going on strike. It was from sound social and liberal principles that Rotterdam saw the launch between 1920 and 1950 of an unprecedented number of public-private initiatives to improve the city. Vreewijk garden village, whose construction

on the south bank near the new harbours began in 1916, offered more than 5,000 working families a home, each with its own garden. Feyenoord Stadium, the Volksuniversiteit (night school) and the Kralingse Bos (woods) would never have come about if it had not been for the commitment of the port elite and their wives to these public works. Besides, far-sighted De Jongh had already laid the groundwork for what could now become an enormous recreation district.

JUMP FROM THE LIFT

'That's taking a chance, that's daring; keep your eye on Vlasblom . . . "Tipped" from the lift-tower, sixty-five meters high', wrote poet Koos Speenhoff in 1933. The occasion was the jump that Lou Vlasblom made on 14 January 1933 from the northern lift-tower of the railway bridge spanning the Koningshaven. With his 65-m dive, he beat the world record by 8 m.

It seemed no end was in sight to the expansion of the port: more and more ships were transferring their cargoes in Rotterdam. A redrawing of the municipal boundary with Schiedam in 1925 made room for the Vierhavengebied or Four Ports District, made up of the three channels of the Merwehaven for the transhipment of mixed cargoes, plus the Vliethaven for petroleum. The construction of the lifting bridge over the Koningshaven in 1927, too, which was nicknamed De Hef, was a feat of engineering prowess. The design was by Pieter Joosting (1867-1942), head of the Bridge Construction Department at Dutch Railways. The prior bridge there, the Koninginnebrug (Queen's Bridge), kept being put out of action by the rigours of traffic, so that locals had dubbed it the Bridge of Sighs. Joosting's design allowed for the entire bridge surface to be raised when tall ships needed to pass.

In the late 1920s, the volatility in the world economy was gradually making itself felt in the port, too. The port was too dependent on transhipment to Germany, which was in the economic doldrums. Classified documents from the 'communist cell' in the port in 1932 reveal that workers at Frans Swarttouw's stevedore firm had to report for business three times a day even though there was seldom any work to be done. The cell's letter to the responsible government minister accuses 'port capitalists' and 'shipping barons' of outsourcing all the costs of the Great Depression to the working man. If the high and mighty were not prepared to listen, the letter threatened, then they would have to feel the effect through strikes.

The setting-up of a Municipal Port Authority (Gemeentelijk Havenbedrijf) in 1932 with its own director, civil engineer Nicolaas Koomans (1894-1970), ensured that the vulnerability of being totally dependent on transhipment was lessened by the city expanding into industrial premises. One of the matters that occasioned its coming into being was entrepreneurs' criticisms of the failure of negotiations with US car manufacturer Ford to build a plant in the city, despite there being thousands of out-of-work dockers at loose ends. In internal management, too, there were shortcomings to be tackled. The Municipal Commercial Establishment lacked a central accounting system, for instance, so that expenses and income were not immediately evident. The Port Authority began coordinating the spatial and logistical organization of the port, too, and earned its keep by imposing tariffs on commercial facilities (especially through berth hire), the piloting service, the docks and the ferries. In 1934, there were already nearly 600 workers in it. The dawn of the new bulwark of the port's economy, petrochemicals, was already apparent before 1940. The Eerste Petroleumhaven was inaugurated in 1933, quickly followed by a second. Shell, Texaco and British Petroleum (BP) were the pioneers of the industry here.

PRINCESS AND PRINCE

Accompanied by the director of the Port Authority N.Th. Koomans and harbour master A. Kortlandt, Princess Juliana and Prince Bernhard wave at the spectators along the quay.

Koomans, remembered for being a director who was 'always sketching away at port plans', stayed at his post until 1959 and ushered the port into the new era. In tandem with its modernizations in the 1930s, Rotterdam City Council and surrounding municipalities unveiled an ambitious plan for the whole IJsselmonde region in 1938, one which strove to strike a balance between port construction and new housing for the city's people. Research had indicated that the left bank of the Maas had burgeoned into a city in its own right in which some 170,000 people shared fewer than five hectares of open space. As there was nothing in the way of recreation, large groups of youths formed what was laconically known as the 'urban proletariat', passing the time in vandalism. It was therefore of the utmost importance that the land usage of this urban landscape be documented through managed spatial planning, to avoid the population degenerating. The growth of the port allowed for new scientific methods of calculating the population's recreational needs; methods that would leave a deep mark on post-war Dutch planning.

Lara Voerman

THE 'PORT OF ROTTERDAM' BRAND

The Havenbedrijf der Gemeente Rotterdam (today known in English as the Port of Rotterdam Authority) was founded in 1932. One of the main aims of both the port sector and the city was to introduce the Port of Rotterdam 'brand' to the market. A foundation established for this purpose, the Stichting Havenbelangen, began focusing in 1933 on the business world and on the port's elite. The foundation organized network meetings – the first Port Day was held in 1935 – and invested in a real lobbyist who would praise the qualities of the port at parties. These kind of 'port offices' already existed in the United States. But Rotterdam's main local rival, Antwerp, also had its own publicity department; when I was on a study trip in England, I learned that many people thought that Rotterdam was 'a Belgian port'.

Before the Second World War, the city and the port formed a single entity, and they supported each other's identities: the tough working city and the economic engine that provided for everyday life. This romantic image of the port was continued during the reconstruction years in the form of port documentaries, major events and artistic photo books. The plans for the harbour basins of the Botlek (starting in 1954), Europoort (1958) and the Maasvlakte (1960s) drove the port westwards, away from the city. Even when standing on the Euromast, peering through binoculars, the port was hard to find. In the 1970s, the port's image began to plummet. Many believed that the city no longer had anything to do with that polluted port (which had since become the largest in the world), and saw it only as source of trouble and nuisance. In the 1970s, the port and the city seemed to be irreconcilable entities, but they did manage to reconcile in the decades that followed. There was good reason for the Open Port Day held in 1979 to choose the theme 'Hello Port'; it allowed Rotterdammers to become reacquainted with 'their' port, and that trend has since continued. Today, the port can again be found in the epicentre of the city, namely on a giant video screen at Rotterdam Central Station. Rotterdammers are invited to the port not only to work, but also for recreation, art and dining.

PART OF ROTTERDAM'S IMAGE 1930 ↓

In the early twentieth century, Rotterdam adopted a rolled-up-sleeves image, derived from the rapid developments in the port area. Rotterdam's strength was represented not by the historical city, but by the cranes and grain elevators, at least according to the city's promotion department. At the Exposition of 1930 in Antwerp, Jaap Gidding's metres-long work of art combined ships, the port and the city into a single overwhelming impression: the 'world's port' as the centre of a vast hinterland.

Jaap Gidding, *Het achterland van Rotterdam,* shown at the Exposition of 1930 in Antwerp, 1930

UNCOMPROMISINGLY MODERN 1933 ↓ →

In the midst of the economic crisis, in 1933 the business community and the newly established Port of Rotterdam Authority joined forces to form the Stichting Havenbelangen, a professional interest group with the vital goal of 'selling our port'. The Rotterdamsche Havenkroniek (Rotterdam Port Chronicle), published by the organisation between 1937 and 1939, was a kind of graphic work of art. The climate of industrial progress had an impact on the city's art and architecture. Companies such as Van Nelle and De Bijenkorf opted for progressive design. New collage techniques, contemporary photography and infographics – which for example depicted the endless fields of storage tanks along the newly excavated Petroleumhavens – connected perfectly with the image of a well-equipped international port.

Rotterdamsche Havenkroniek, no. 2 (1938)

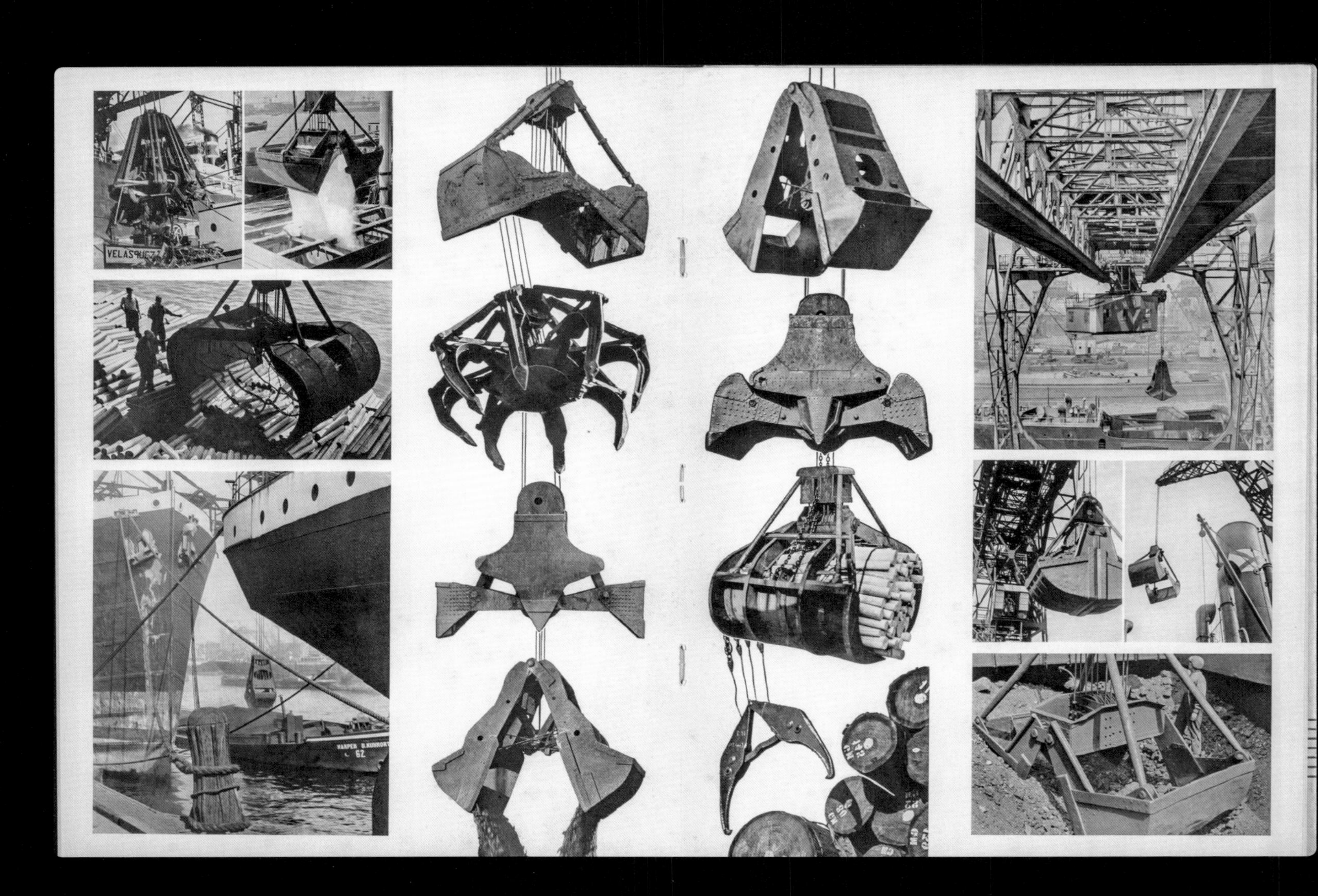

As a centre for the import, manufacture and distribution of mineral oil products, the Port of Rotterdam has growing significance. Diagram of the imports in the period from 1925 to 1937

The work in the port has been a source of inspiration for numbers of artists, painters and etchers as well as photographers and camera-men.

Photos from the film "The port of Rotterdam" by A. v. Barsy.

Photographs: A. v. Barsy, C. Kierdorff, C. Kramer (N.F.P.V.), J. v. Maanen, H. A. v. Oudgaarden (N.F.P.V.) W. Schack, F. Schneider, H. v. d. Winkel, etc. - Air pictures: K.L.M.

8 ←

0s avant-garde filmmaker and photo-
med films about the port, including
k (1938). In these films, the port was
es, tugboats and port workers, while
ground. With these films, Von Barsy
l after the war.

l of the port's reconstruction period.
completed successfully: in 1950 there
en in 1938. The city celebrated this
d was over. With more than 1.5 million
1950s for both Rotterdam and the

THE PORT IN MINIATURE 1952 ↓ →

Madurodam was designed in 1952 as a Dutch miniature city on a scale of 1:25. It was an excellent opportunity for companies to present themselves to the public. KLM donated an airport, the NS donated a railway system and Philips provided the lighting. Port of Rotterdam Authority donated a fully equipped seaport, the Petroleum Maatschappijen donated a petroleum port with tanks and a drilling platform, and several Rotterdam shipping companies donated large ships. Even today, the port is still an indispensable part of Madurodam, and the miniatures of the port have since been realistically expanded to include the Maasvlakte 2.

The Port of Rotterdam in Madurodam, The Hague

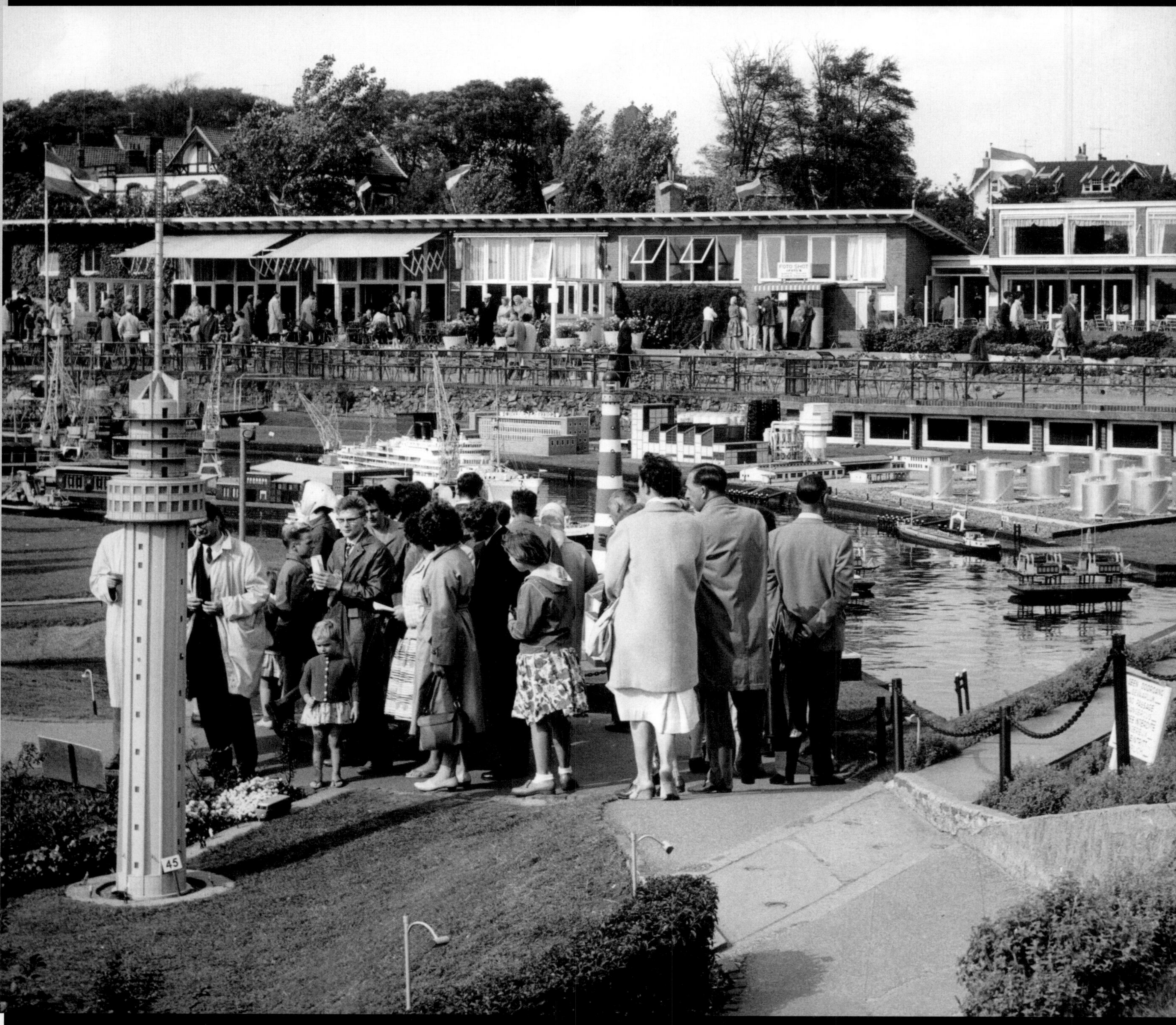

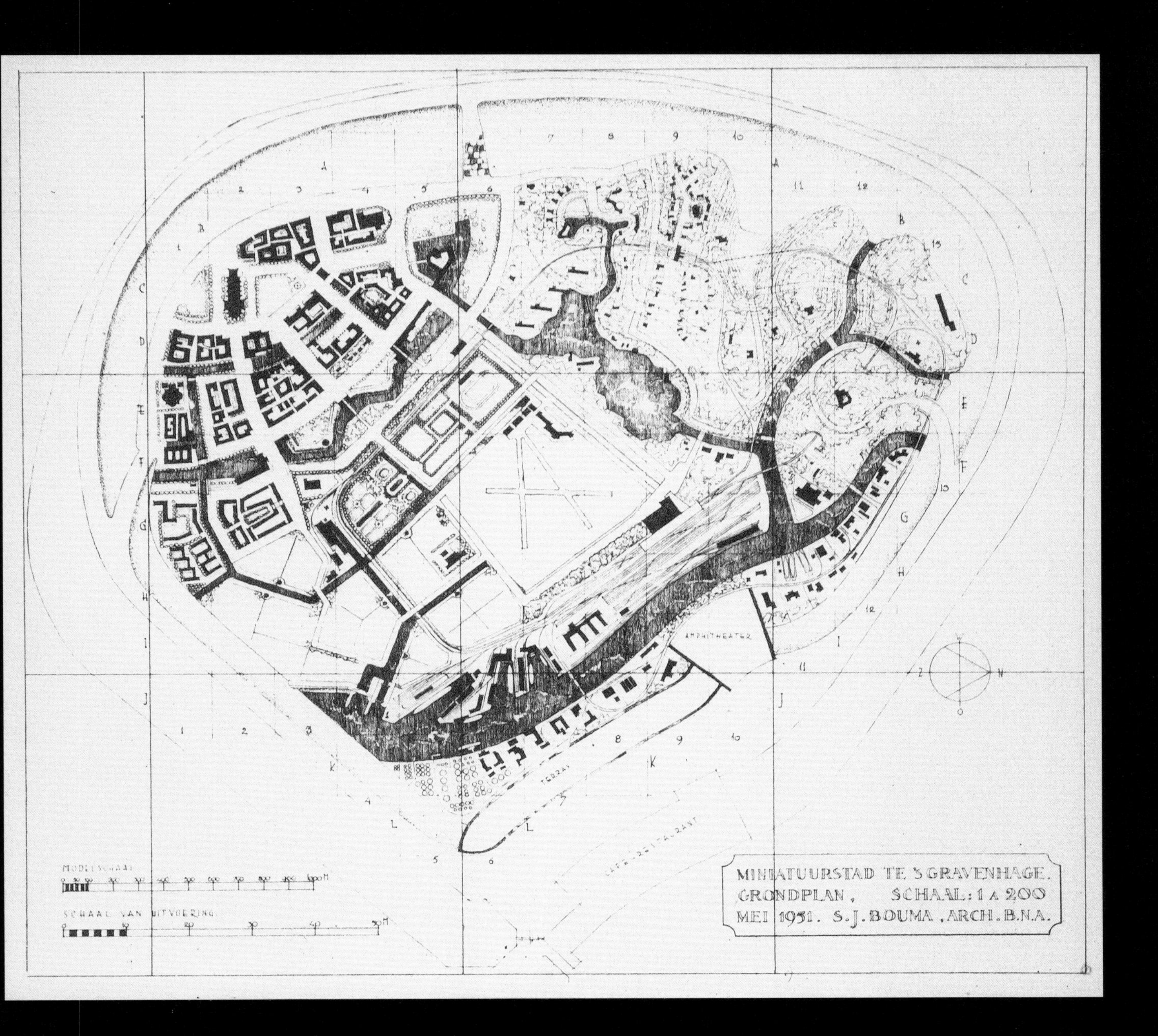
AMPHITHEATER
MODELSCHAAL
SCHAAL VAN UITVOERING
MINIATUURSTAD TE 'S GRAVENHAGE.
GRONDPLAN, SCHAAL: 1 A 200
MEI 1951. S.J. BOUMA, ARCH. B.N.A.

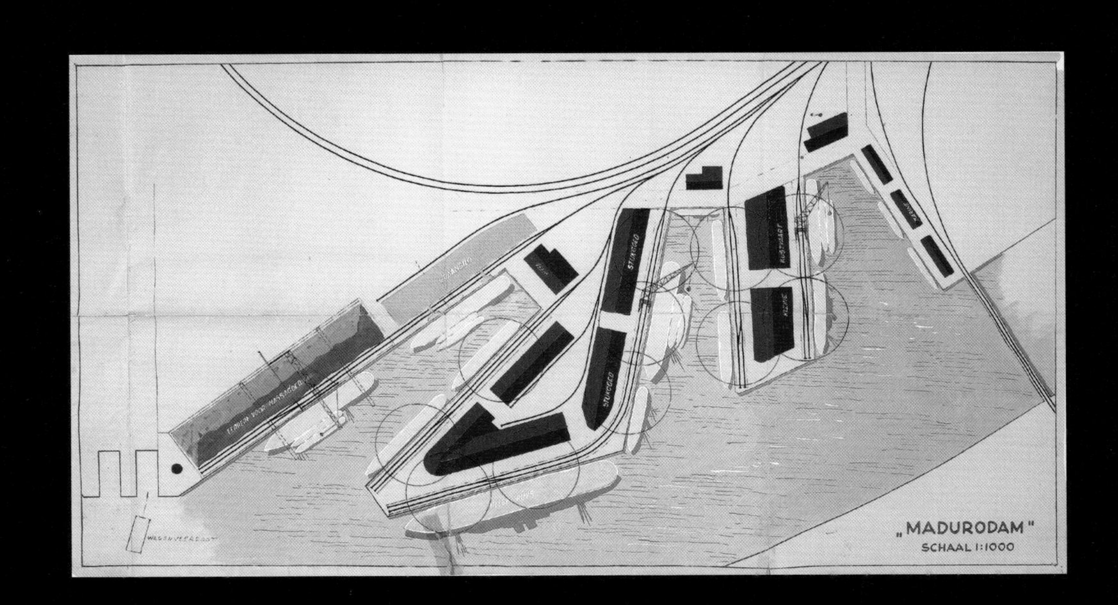
„MADURODAM"
SCHAAL 1:1000

THE NOSTALGIC PORT 1955 ↓

'Steam winches rattle, electric tools whir, engines hum and pop. Cranes, long-necked like giraffes, sway and bend over huge ships and double rows of caissons. Huge floating cranes, true goliaths, torsos of heavy pieces of machinery.' In 1955 the Stichting Havenbelangen published *Rotterdam. Safely Moored, Quick Dispatch, World Port*. The modern message in this title is not consistent with the lyrical texts and romantic sketches and photographs that are contained in this small book. While the port was growing exponentially, there seemed to be a nostalgia for the heroic image of the port before the war: a port with a human dimension.

Rotterdam. Safely Moored, Quick Despatch, World Port (Rotterdam: Stichting Havenbelangen, 1955), 34

MONUMENT TO A TRANSITION 1958 →

In 1958, the go-ahead was given for the grand plan known as Europoort: a series of new harbour basins located several kilometres to the west. The city and the port were drifting apart. As a final attempt to reinforce the relationship between the two, a huge 'mast' was installed at the Floriade event in Het Park. The name Euromast was coined by the director of the Port Authority. Only on the viewing platform, which architect Hugh Maaskant designed with good reason as a ship's bridge with a wheelhouse, could a glimpse of the working harbour be had through binoculars.

The Euromast, 1966

HANNO
FIGEE
TRAUGUTT

THE PORT OF EUROPE 1962 ↓

The construction of the Europoort meant that in 1962, the Port of Rotterdam had become the largest port in the world. For the first time, the downside of Rotterdam's urge to modernize and expand was exposed, for example in the television documentary *Polders voor industrie*, which showed how nature reserves, villages and farms had to disappear in order to make way for the oil refining and chemical industries. In the 1960s, the Port Authority commissioned prominent filmmakers to revamp the port's fading image. In 1962, Ytzen Brusse filmed *Poort van Europa*, Joris Ivens made *Rotterdam Europoort* in 1966, and Tom Tholen made *Toets* in 1967.

'Zeg Weena, zeg Spangen, zeg Kralingen, Crooswijk, zeg Feijenoord, Dijkzigt, zeg Oud-Mathenesse, zeg honderden namen, zeg stad aan de Maas, zeg Poort van Europa', *Rotterdam Europoort*, 3 (1966) 1

Stills from the VPRO documentary *Polders voor industrie*, 1962 (directed by Wim van der Velde)
New construction in the village of Rozenburg, showing the Botlek in the background

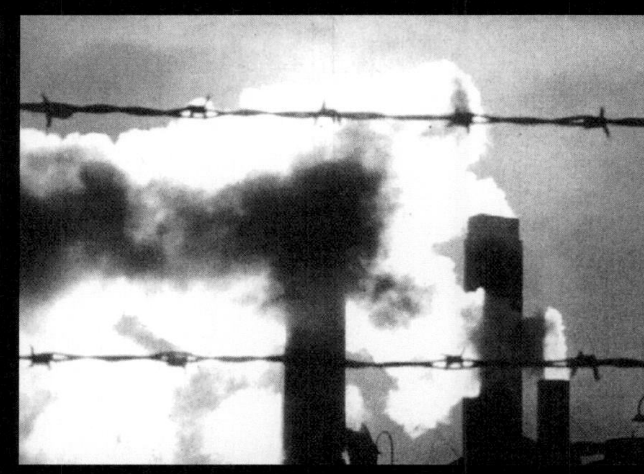

TOO MUCH, TOO FAST 1973 ↓

In the early 1970s, the newly opened Maasvlakte, supertankers and the chemical industry gave the port the image of being and creating a nuisance. The port was suddenly viewed with a critical eye by both politics (a new left-leaning government) and by the city itself (demonstrations on the Coolsingel against the arrival of new blast furnaces). The publication of the famous report *The Limits to Growth* by the Club of Rome in 1972, as well as the 1973 oil crisis, contributed to the growing distrust of this 'dirty' industrial world beyond the city. Slowly and sparsely, the promotional magazine *Rotterdam Europoort* began publishing its first articles about nature and the environment.

Rotterdam Europoort, 3 (1966) 1, (cover)
Rotterdam Europoort, 5 (1968) 3, 18

Beplanting in het Europoort-Maasvlaktegebied

De rondritten door Botlek en Europoort, die het Rotterdams Vervoersbedrijf in zijn zomerprogramma opnam, kwamen vroeger sterk overeen met de excursies door de woestijn naar de piramiden in Egypte. Beide landschappen imponerend door hun kale grootsheid met hier en daar een machtig bouwwerk. Daar ter ering van dode koningen, hier voor het dagelijks brood van levende mensen.
Het was toen ook reeds interessant om enkele uren in deze grootse omgeving te vertoeven, maar wel eens deprimerend voor die mensen van het dagelijks brood.
Geen woongebied doch een werkgebied, maar ook hiervoor te kil, te hard en te ongevoelig.

In 1960 werd besloten om door een groenbeplanting van het haven- en industriegebied het werkklimaat van dit gebied te verbeteren.

Een groenaanleg in zo'n bar gebied ontmoet welhaast alle moeilijkheden die aan een beplanting in de weg gelegd kunnen worden.
Een bijzonder onvruchtbare bodem die ongeschikt is voor plantengroei; wind met grote mechanische kracht; wind beladen met zeezout; wind beladen met stuifzand en ten slotte een keiharde groep werkers in de cabines van draglines, tractors en vrachtauto's, die in dit gebied in volle opbouw hun sporen door het land trekken, die meer gewend zijn in het zand te rijden dan op een gebaande weg en die door hun zijruitjes, als daar horizontaal de regen tegenaan klettert, geen aandacht schenken aan een paar kale takjes van een pril heestertje langs de weg, dat een tuinman daar zorgvuldig geplant heeft.

En toch is er in vijf jaren al heel wat opgegroeid, omdat geen van de genoemde moeilijkheden onoverkomelijk waren.

HELLO PORT 1979 ↓

In the 1970s, the port and the city seemed to be incompatible. The well-attended event *Rotterdam Maratiem '78* led to the idea of an 'open day' at the port, which took place for the first time the following year. Rotterdam's Havendag (Port Day) had existed since 1935 as a networking event for port magnates, captains of industry and politicians, but this concept had become outdated. The theme of the first Open Port Day, 'Hello Port', was a significant choice. However far away the port might be, it had to be brought closer to the people of Rotterdam. A special 'Europoort express' train rode along the port's railway line, and offered new views of the impressive harbours and industrial sites. Since 1979, the event has grown into a three-day manifestation, attracting hundreds of thousands of visitors.

Wereldhavendagen (World Port Days), year unknown

BETTER INSTEAD OF BIGGER 1980 ↓

In the 1970s, the fact was consistently stressed that Rotterdam was the largest port. But after a poorly received new port film – critics called it a 'television advert' with an 'aggressive American voiceover' – the Stichting Havenbelangen took a different tack in the 1980s. With the female president Joyce Bosman-Kater at the helm, the arrogance of power disappeared. In a period when the flow of goods to and from Rotterdam had stabilized, the message changed. Instead of being the biggest, Rotterdam would be the best: reliable, well-equipped, fast, safe and skilful.

Rotterdam Europoort Delta, 17 (1979) 1, 3

Europe... the better gateway

Rotterdam/Europoort is chosen by more shippers than any other port. We handle an average of 33,000 ocean-going vessels annually, carrying 275-million tons of cargo.
Why do so many customers select us? Not on mere promises. Rather because we do what we say. This we offer:
Speed: Quickest possible ship turn-rounds. Fast intra-Europe transportation by road, rail, air, inland barges and pipeline.
Efficiency: The most modern facilities and equipment needed by container ships, conventional vessels, tankers, and any other type of carrier.
Skill: Dependable personnel trained in specialized schools.
Connections: 12,500 Regular sailings a year offer a profusion of worldwide services very hard to beat.
Continuity: Proud record of good labor relations making Rotterdam a reliable distribution center.
And...
No other port in Europe is better located - right where the Rhine meets the sea. No locks, no bridges. The center of a 160-million people market - producing and buying.
No other port can serve Europe the way Rotterdam can.
The busiest port of them all.
Because Rotterdam/Europoort cares most... about service.

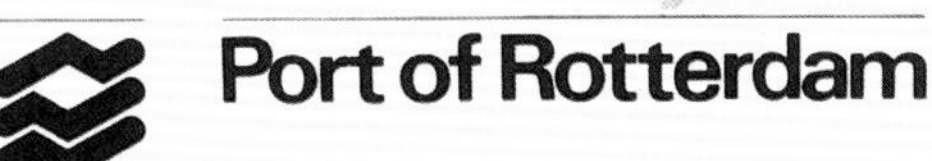

Port of Rotterdam - External Relations Department - 'Europoint III' - Galvanistraat 15 - 3029 AD Rotterdam
Phone: (010) 89 69 11 - Telex 23077 - Cables: 'eurogate rotterdam'
Postal address: P.O. Box 6622 - 3002 AP Rotterdam

ROTTERDAM MAINPORT 1985 ↓

In the early 1980s, the Port Authority lobbied to transform the port into a mainport: the main hub for goods in Europe. In order to achieve this, it became clear that investing in only the port would not be enough; the city of Rotterdam had to change along with its port. The port hereby challenged the city, and did so successfully. Rotterdam began investing in high-quality housing areas and in space for innovative businesses, culture and education. The idea of Rotterdam Mainport is the basis for the redevelopment of areas such as Kop van Zuid, the Müller and Lloyd piers and the RDM campus: places where the port could again be felt in the city.

Wilhelminahof along the Kop van Zuid, under construction in May 1995

MAKE IT HAPPEN 2014 ↓

Much like in the 1930s, in 2014 the port and the city again combined forces. Under the slogan 'Make It Happen', they profiled themselves by using each other's strengths: a pioneering spirit, dynamism and knowledge. With the largest video screen in Europe, the Port Authority lets passengers in the main hall of Rotterdam Central Station know that they have arrived at Europe's largest port. The Port Authority encourages the public to travel to the port for other activities, through initiatives such as Futureland (the information centre about the Maasvlakte), the construction of new beaches and a network of bicycle paths that spans 120 km. It also sponsors icons such as De Doelen, Rotterdam Zoo and the North Sea Jazz Festival.

The Port Authority provides the content for the large video screen in the station's main hall.

PORT PLACES

Frank de Kruif [FdK]
Isabelle Vries [IV]
Peter Paul Witsen [PPW]

photography
Jannes Linders
Siebe Swart

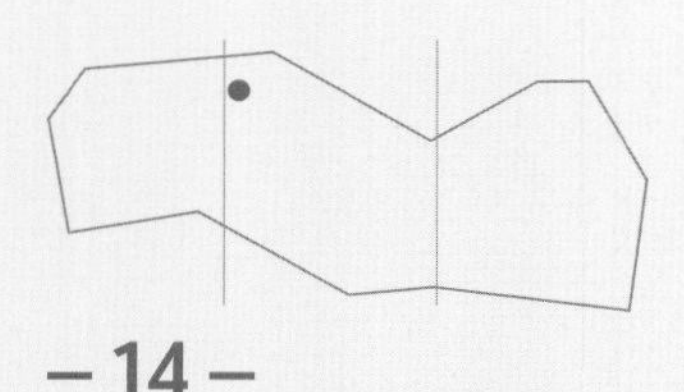

– 14 –

Bagijneland 11 \ 's-Gravenzande

WESTLAND

Cycles

The finite nature of resources and the increasing sea level make us aware of the importance of the 'circular economy'. This concept precisely indicates that there is an economic justification to be striven for here. The circular economy focuses on reusing waste, products and raw materials as much as possible, and thereby minimizing the destruction of value.

Ideas about the circular economy are not new. The shining example in the world of industry has long been the Danish town of Kalundborg. Under the rubric of 'industrial ecology', a number of companies settled here in the 1970s; these firms worked together, mutually exchanging residual by-products and thereby saving on raw materials. The connection between the companies was initially motivated by economic incentives, but gradually they became aware of the positive effects that this was having on the environment.

In the Port of Rotterdam, companies have been exchanging a diverse range of materials for many decades. And the heat generated by the waste incineration goes to downtown Rotterdam. In Westland, the CO_2 that is released by industrial processes is used for crop cultivation. And all of this takes place underground, invisibly.

The phenomenon of 'exchange' became visible to the public at the port when urban agriculture (*avant la lettre*) arrived there. It was the Happy Shrimp Farm, a unique nursery for tropical shrimp that opened its doors in 2006. This innovative company used waste heat from a power plant on the Maasvlakte for breeding the shrimp. In its short existence, the Happy Shrimp Farm received widespread me dia attention and visits from enthusiastic politicians and won many prizes. Renowned fish shops and restaurants in the city put their shrimp on the menus under the name Happy Shrimps. Unfortunately, there were delays in production and high start-up losses, and the young company was forced to close its greenhouses in 2008. Perhaps the port was not yet ready for these kind of start-ups.

The Happy Shrimp Farm was seen as a bit silly, but since then the reuse of heat, steam and CO_2 has become regarded as *the* major challenge for the Port of Rotterdam. A Delta Plan is being developed for the entire region. The heat from the port complex can be used to heat other businesses, homes or greenhouses. Talk about extra value.

Things will only become truly circular when we are able to produce fuels and other consumer goods from sustainable, renewable sources. The port, which so successfully capitalized on the rise of coal, oil and gas, will also have to make the transition to green resources.

– IV –

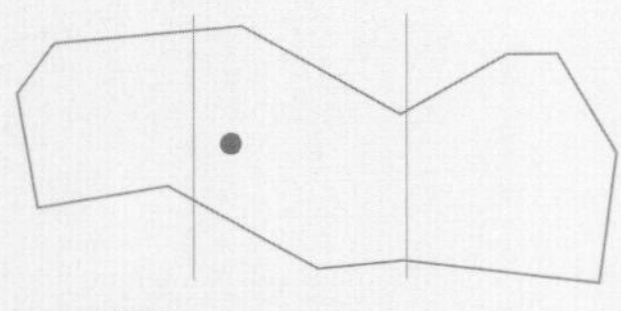

— 15 —

Merwedeweg 5629 \ Europoort Rotterdam

EUROPOORT

Green Chemistry

Is biomass the green gold, the new mainstay that will soon be driving the port? From fossil port to bioport is a logical step in combating climate change and the depletion of the earth. With its infrastructure, industry and space, Rotterdam is of course in an excellent position in this regard. But until now, this has not turned out to be so easy, and the big breakthrough remains elusive. The oil price is still too low, as is the price of CO_2. Shale gas is being extracted and people are frantically searching for oil in the fragile Arctic. Biomass is often still too expensive, the market is uncertain and policy vacillates too much. Subsidies for driving on bioethanol, for example, have already been eliminated.

The green economy also poses ethical issues. Biomass, which consists of plant debris, can be used for energy production, production of biofuels and bioplastics. But the right biomass is not always available in large quantities. We prefer to keep raw materials that are suitable for food in the food chain. We don't cut down jungles for the purpose of extracting palm oil, for example.

Producing sufficient, ethically responsible feedstock would have to be done on such a large scale that it would come at the expense of lots of farmland. Experts have calculated that at our current consumption patterns and energy needs, it would be impossible for the world to switch to 100 per cent green raw materials. That means that energy conservation and the use of renewable energy sources such as wind, solar or hydrogen remain vitally important.

In bio-based chemistry, the trick is to optimize the use of all parts of the plant as much as possible, so that they yield the highest return and create the least waste. That means using high-calorific components for medicine and food, using the remaining components for chemical products and only what's left over for biofuels. Finally, the non-usable residues can be stoked in power stations. Process technology will have to be developed much further. This offers excellent opportunities for knowledge development and innovation in the Rotterdam region.

It might still take a generation or two, but eventually there will be a transition. All in all, it would of course be fantastic if we could replace our addiction to oil and plastic with an addiction to plants.

– IV –

VAN DE LAGEMAAT
TRANSPORT
SCANIA
SUPER
R 420
BV-BF-52

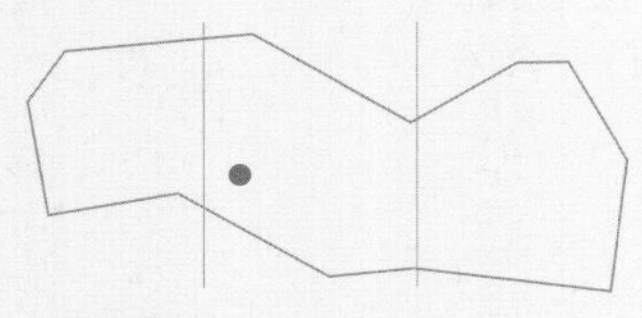

– 16 –

Brielse Veerweg 3 \ Brielle

BRIELSE MEER

24-hour Water Machine

Mobile homes, seemingly placed at random among the trees. Wide sandy plains between bushes. Cormorants on stone piers. A nudist camping site. The shore of the Brielse Meer (Lake Brielle) is not a place of great scenic beauty, but it does provide Rotterdam with a pleasant recreation area.

Even before the flooding of 1953, the Brielse Maas was dammed, which created the Brielsemeer. This was not for security reasons, nor to make room for the port or for mobile homes. The reason was that the agriculture and horticulture of Voorne-Putten and Rozenburg (which was then still an agrarian town) needed fresh water. The Brielse Maas was brackish, and the salty seawater flowed far into the river and penetrated the groundwater. The construction of the dam in the Brielle Maas instantly decreased the salty shoreline by about 50 km. As a shipping route, the river could be missed: the New Waterway offered an adequate alternative, followed not much later by the Hartelkanaal. Much in the same way that the Oostvoornsemeer on the other side of the dam was kept salty, the Brielsemeer is meant to be a freshwater lake.

The Brielsemeer is now the main fresh-water reservoir in the region. It is used not only for agriculture and horticulture at Voorne-Putten, but also for Delfland: the Winsemius pumping station pumps water into a pipe that travels under the Europoort and the New Waterway, on its way to Westland. The port and industrial complex also make use of this water. The petrochemical industry, for example, needs demineralized (super clean) water. All told, for the industrial uses plus agriculture and horticulture, this amounts to an average of about 4,000 litres per second. Almost all of that water comes

from the docile stream called the Bernisse, which at Heenvliet flows into the supply channel for the Brielse Meer, and is in turn fed by the river.

The pleasant, somewhat sleepy recreation paradise of the Brielse Meer turns out to be the centrepiece of a water machine that is almost always running at full speed, and that is vital for the industry. If the supply of fresh water were to stop, this machine would grind to a halt within a matter of days. At the same time, the risk of the Bernisse becoming salinized is increasing, in part because several dozen kilometres away, fresh seawater will soon be allowed to flow into the Volkerak and the Zoommeer in order to solve problems with the water quality there. In cases of storms or drought, saltwater can flow into the Bernisse. Methods are being sought to make the Brielsemeer less dependent on this single source, for example by restoring an old inlet point at Spijkenisse. This simultaneously shows the ingenuity and the vulnerability of the water system across the entire river delta – invisible to those who are looking for relaxation on a camping site along the rippling lake.

– PPW –

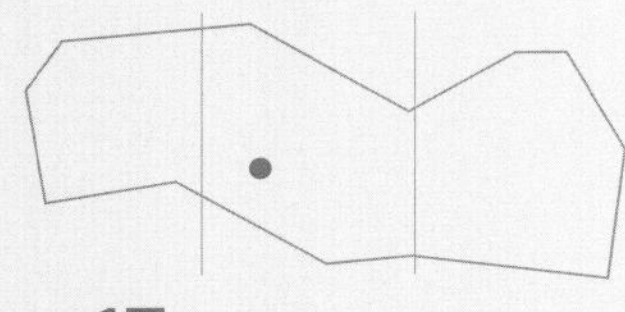

– 17 –

Moezelweg 75 \ Europoort Rotterdam

7^E PETROLEUMHAVEN

Information for Everyone

Developments in the field of data and information technology provide unprecedented opportunities for the port. Examples include giving the owner or shipper insight into where a container is located, and indicating when it will arrive at its destination. Or making better use of infrastructure so that you can track all your vehicles and provide them with real-time information. Or drones in the air or underwater to carry out inspections on pipeline corridors or ships.

An example of the 'Internet of things' (linking the physical world to the digital world) are the sniffing noses in the port area. But you have to know that they are there to be able to see them, because they disappear amid the massiveness of the port's infrastructure. These electronic noses detect changes in the smell composition of the air. Gases that are detected have a specific pattern, a kind of barcode. The e-nose compares this to all the known patterns in its digital library. New, unknown patterns are added to this database, which means that the noses are able to learn. The e-nose gives information about the location, type and intensity of the gases, and sends this information to the incident room of the DCMR. In cases of dramatic change, the company causing the gases is asked to take action and the local residents are informed. As a co-initiator of the project, Vopak has these noses on its own site. This allows the company to take the correct measures on its own, and in time, before nearby residents are inconvenienced.

The age in which residents themselves will be able to control the quality of their living environment is not far off. They are demanding – and getting – access to data and information. If that data and information are not provided, they take the measurements themselves and share this data with the neighbourhood via Facebook or WhatsApp. Students from the Hogeschool Rotterdam have already developed an e-nose for bicycles.

Big data and open data are now seen as a very decisive factor in the success of the port. 'Position yourself at the heart of the logistics Internet', Rifkin told the Port Authority. Making use of data in the port is very logical, but actually allowing this to happen is complex. The various parties need to be transparent, but they often see information as their property. They do not like to offer evaluations of their own performance. And there is no single authority or editor who makes decisions or determinations; the existing hierarchical structures no longer work. The proper use of data in the port requires different forms of cooperation and leadership, especially leadership in making connections.

– IV –

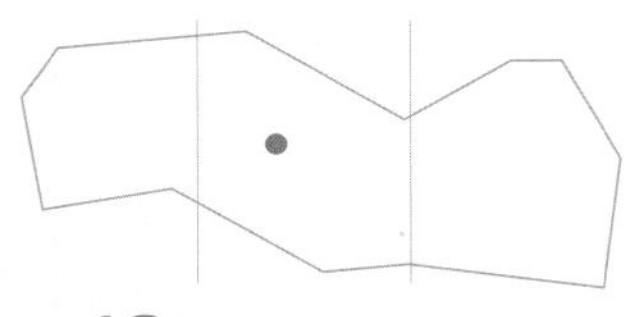

– 18 –

Merellaan 943-1385 \ Maassluis

MAASSLUIS

Balcony with a Seaport Exemption

The sun shines into the house through the French balcony, a glass of white wine is at hand, and ships glide by far below. This sums up the dream of living in a world port, and the Waterwegtorens (Waterway Towers) appeal to that dream. It consists of three brazen residential towers in Maassluis, up to 21 storeys high, just behind the dike. From a distance, they seem to merge into one massive building volume. They were built in 2007 on the site of a number of obsolete gallery flats.

But living in a world port also means that you can hear the hum of generators on moored vessels. Sometimes plumes of industrial smoke blow through the French balcony into the house. The Waterwegtorens were constructed under a special legal status: the seaport exemption. The idea was that the thrill of living in a special residential environment would allow for a few extra decibels of industrial noise. The private sector was given the assurance that it would not have to satisfy tougher environmental requirements as a result of the new building. For the most recent housing project on the banks of the Maas in Maassluis, called Het Balkon (The Balcony), no special regulation was required. This area is not located directly across from the industrial area, but is instead across from the island of Rozenburg. That means that the limits of environmental hinder for that building are the same as they are for the rest of the Netherlands. Companies in the port have their doubts about these kinds of residential complexes. They fear that residents will be quick to complain to the regional environmental agency DCMR, even if the companies have not violated any of the environmental norms or permits. Companies in the port have their doubts about residential complexes like these. They are afraid that residents will soon complain to the regional environmental agency DCMR, even though everything is properly within the limits in terms of standards and permits.

Most of the complaints that the DCMR has received about the port and its related industries have to do with noise or odours. There are a few thousand such complaints each year, spread out across the entire region. Most of the complaints can be traced back to incidents such as faulty equipment or special activities, or exceptional weather conditions.

Some companies take an extra step, and try to keep ahead of the incidents that will cause complaints. The sidewalk at the Waterwegtorens, for example, has an e-nose. Maassluis has four of these e-noses, two of them were donated by the tank storage company Vopak.

This shows how the industries of the port are looking for a new relationship with the people who live there. They make contributions to the quality of the residential environment, but at the same time they don't want that residential environment to get too close. Because a resident who is driven from his balcony by odours or noise will not care about environmental contours or legal exemptions. He will file a complaint.

– PPW –

In the foreground Het Balkon, with the Waterwegtorens to the right in the background.

JUTLANDIA SEAWAYS
LONDON
IMO NO. 9395355

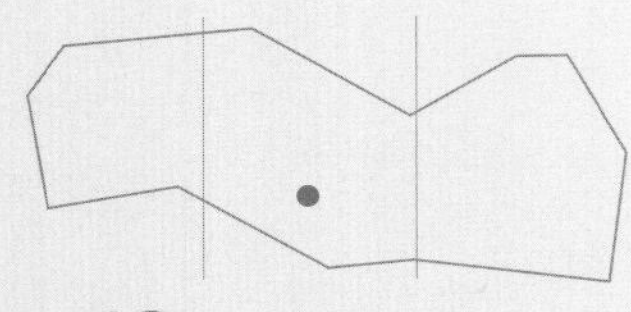

– 19 –

Merseyweg 50 \ Botlek Rotterdam

BRITTANIËHAVEN

Dock for Floating Parking Garages

It is the boyhood dream of many a car lover: driving brand new cars from a ship to the quay, all day long. But be careful, of course, because bodywork damage results in significant costs to the Rotterdam Car Terminal in the Botlek.

Many newly built cars sail the world's seas aboard ro/ro ships: *roll on, roll off*. Today's largest car carriers can transport 8,000 vehicles. These sailing parking garages are not the most stable vessels, and they occasionally capsize. In 2003, the *Tricolor* sank in the Channel, with nearly 3,000 brand-new BMWs, Volvos and Saabs on board. Some are still in the Channel.

But usually, the ships safely deliver these cars. Or anything else that rolls: trucks, tractors, bulldozers or tanks, as long as they can be driven. The Rotterdam Car Terminal mainly handles new cars. Rotterdam, with about 200,000 vehicles per year, is certainly no leader in this field; competitors such as Bremerhaven and Zeebrugge handle ten times as many.

The car terminal in Brittanniëhaven used to be part of ECT, but this container company neglected the development of ro/ro transhipment. After the acquisition by Broekman, the terminal rebounded, which was reflected in the construction of four car parks. Located right next to the A15, they can't be missed.

What Broekman primarily invested in was providing added value: installing radios and navigation systems, mounting special wheels and adding instruction manuals in the appropriate language. These are all tasks that the importers have chosen, for whatever reason, to outsource.

Broekman has since sold the terminal. The new owner is its old neighbour Cobelfret, the Belgian company that already operates a ro/ro terminal in that same Brittanniëhaven.

– FdK –

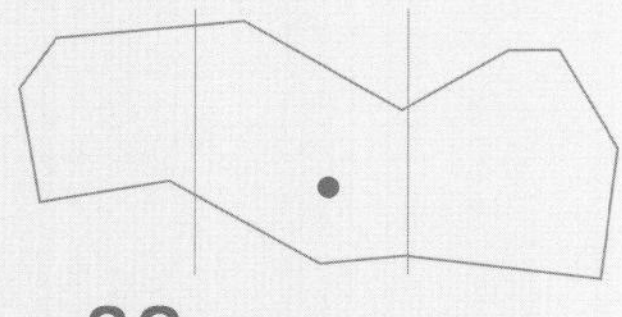

– 20 –

Prof. Gerbrandyweg 17 \ Botlek Rotterdam

BOTLEK

Storage for Futures Markets

Steinweg is one of the oldest and most successful companies in the Port of Rotterdam, but it is not the most famous. That has to do with the market in which storage and transhipment operates: international trade in raw materials, the so-called commodities. Examples include metals such as copper and aluminium, or agricultural products like coffee and cocoa, but also plastics. What these materials have in common is that, just like stocks, they are traded on exchanges.

Commodity trading is a complicated game in which parties enter into contracts to deliver an amount of goods at a specific time at an agreed-upon price. But by the time the contract expires, the price of the goods on the exchange may have increased or decreased. In the former case, this will benefit the buyer, in the latter case the seller. But often at the end of the period, the goods are not physically delivered, but instead the contracts are traded. This happens because the trade in raw materials with fluctuating prices can be lucrative, again much like shares. Many banks, trading firms, hedge funds and speculators have made a specialty of this game.

Just as a shareholder is a part-owner of a real, existing business, the contract holder is the owner of an actual batch of raw materials. The ownership of this batch can often change hands without the goods actually ever moving. The goods are stored for a long time, sometimes for years, and

usually at a port, because the large amounts of raw materials will eventually be transported by ship. And the storage does not take place just anywhere in a port, but in the warehouses of a select number of companies that have obtained a license from the exchanges. These companies, of course, receive a fee for the storage.

Steinweg is one of the companies that has these licenses, for example from the London Metal Exchange and the London International Financial Futures and Options Exchange. Similar companies include Henry Bath and Pacorini, which are also not among the port's most famous names. The reason that these companies prefer to work in the shadows is that publicity about the am

– FdK –

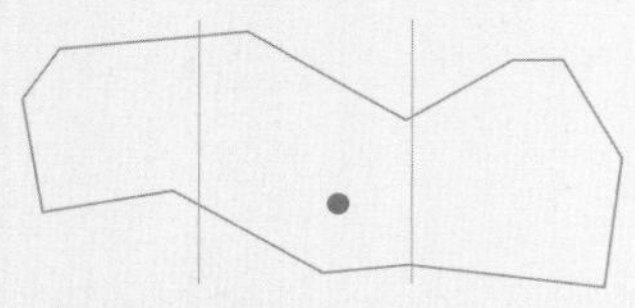

– 21 –

Prof. Gerbrandyweg 25 \ Botlek Rotterdam

DOK 7

Restart of the Shipbuilding Industry

On Dok 7, the mammoth dock of Keppel Verolme in the Botlek, is a special vessel for extracting oil and gas at sea. The *Armada Intrepid* is being prepared there for transport to the Far East.

The Netherlands was once a leading shipbuilding country. Perhaps it still is, but not in the way that it was in 1970, when Cornelis Verolme constructed this dock (then and now one of the largest dry docks in the world) for the purpose of building tankers and other ships. But only a few years later, the first oil crisis arrived, which meant less demand for supertankers; the construction of new ships then became an unprofitable activity.

Verolme, Wilton-Fijenoord, the Rotterdamse Droogdok Maatschappij (RDM): illustrious names from Rotterdam's rich shipbuilding history, names that live on in the places in the port that they once occupied. New constructions have been replaced by maintenance and repair companies, not only for vessels but also for offshore installations. The manufacturing industry still plays as important a role as ever in the economy of the port.

Verolme became Keppel Verolme, Wilton-Fijenoord is now Damen Shiprepair and RDM now stands for 'Research, Design & Manufacturing'; it is an incubator for small businesses that will hopefully strengthen the position of Rotterdam in the maritime manufacturing industry.

The fact that this is a capable industry sector is proven by the operations involving the *Armada Intrepid*. After this FPSO (Floating Production, Storage and Off-loading) vessel was lowered back into the water, it was towed to the offshore facility in the Caland Channel, where it was loaded on the *Dockwise Vanguard*, the largest heavy-lift ship in the world. This is also a form of Dutch glory, because the original Dutch shipping company Dockwise is owned by Boskalis.

The *Dockwise Vanguard* is a semi-submersible ship. It can be submerged so that the object that needs to be transported can sail over it. If things go well, it floats itself back to the surface, and the load is then secured. This is also how the 245-m-long *Armada Intrepid*, which weighs 60,000 tonnes, was moved. Now it is sailing onwards to the Far East.

– FdK –

DOCKWISE
VANGUARD
42M
41M
40M
39M
38M
37M
36M
35M
34M
33M
32M
31M
30M
29M
28M
27M
26M
25M
24M
23M
22M
21M
20M
DOCKWISE VANGUARD

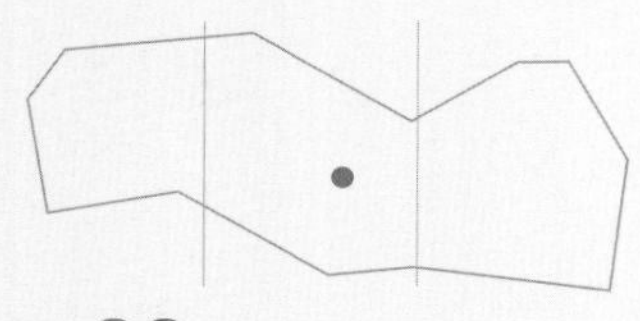

– 22 –

Zuidbuurt 40 \ Vlaardingen

VOLKSBOS VLAARDINGEN

A Planning Controversy

In the open polder landscape between Vlaardingen and Maassluis, next to a winery, a new forest has been planted, with 1,700 young trees and shrubs. This planting was not the result of a development plan for nature or the landscape, as is usually the case with new nature areas in the Netherlands. On the contrary, this forest is a response to the proposed construction of a new highway between the port and the Westland, under the New Waterway through the Blankenburg Tunnel. The local action committee known as Groeiend Verzet (Growing Resistance) planted this 'People's Forest' in an attempt to prevent that plan from being implemented.

The large scale of the infrastructure projects related to the port almost always raise hackles. The Blankenburg connection is no exception. On one hand she is backed by the government, with the municipality of Rotterdam on her side. The connection is necessary, they argue, because the main road around Rotterdam is threatening to become full, and this is reducing Rotterdam's competitiveness. Moreover, the only form of access, via the A15, is vulnerable; one calamity and work in the port would grind to a halt. The expanding relationship with the Westland area justifies a direct connection.

On the other side, the Actiecomité Blankenburg Tunnel Nee (Action Committee No Blankenburg Tunnel), affiliated with Groeiend Verzet, is supported by the Vereniging Natuurmonumenten and other nature and landscape organizations. They point to the irreversible degradation of the landscape, nature and the environment. This area is part of the recreationally attractive and historically valuable cultural landscape of Midden-Delfland. They dispute the traffic forecasts, arguing that the trend towards ever-increasing traffic flows has already stopped, or has at least decreased.

In the Netherlands, planning controversies like these are fought out in the form of reports and counter-reports, and settled on boardroom tables. These cases will then often end up at the administrative court of the Council of State, as opponents challenge the lawfulness of the decision.

Groeiend Verzet carries out its protests by developing nature, and this has proven to be a successful formula. In the 1990s, the foundation was able to stop a landfill for industrial waste by planting trees. That original 'Volksbos' is now managed by the Recreatieschap Midden-Delfland. Whether the second Volksbos will also get the chance to grow remains to be seen. It is likely that the formal decision about the new highway will be made in 2016. The new highway will be built as a sunken road, and will be covered where it intersects with the Volksbos. There are funds of 25 million euros available to take additional measures for nature, green space, water, recreation and soundproofing. The minister of Infrastructure and the Environment hopes that cars will be able to start using the new road in 2022.

– PPW –

On the other side is Midden-Delftland, with the Volksbos.

The former Volksbos.

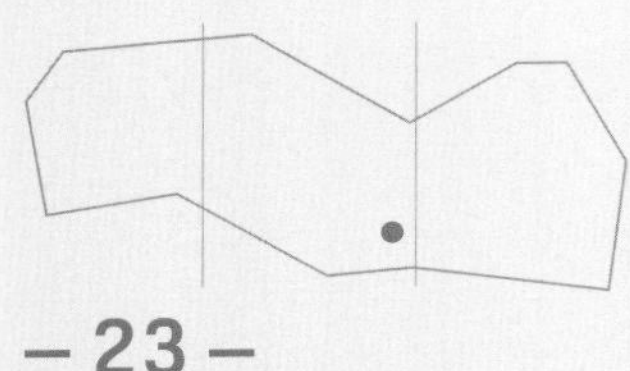

– 23 –

Herikweg 5 \ Hoogvliet

HOOGVLIET

The Port in the Back Garden

The rise of the petrochemical industry around the Second World War caused a planning problem. It became clear that the personnel could not live at the industrial complex itself, which was still possible in 1914, when the village of Heijplaat was built for the workers of the Rotterdamsche Droogdok Maatschappij. Since then, there had been increased concern for the quality of residential neighbourhoods – the ideas of the garden city were developing rapidly.

The Bataafsche Petroleum Maatschappij (a subsidiary of Shell) built its plant on the left bank of the Maas, in Vlaardingen. The Waalhaven and the Eemhaven separated these first two Petroleumhavens (oil ports) from the rest of Rotterdam. But there was no underground metro yet, and a private car was still reserved for a small, affluent part of the population. A residential area was needed for personnel, within cycling distance of the refinery. The location for this new residential area was found around the dike village of Hoogvliet.

Immediately after liberation, the municipal urban planners started drawing. The new Hoogvliet, prompted by the post-war housing shortage, was much larger than Shell needed it to be. The sketches followed each other in rapid succession, but they all shared one element in common: a spacious green structure surrounded by a green buffer. In the north, this created a 'soft' transition to the petrochemical complex. The workers and other residents were able to enjoy a green, almost idyllic environment. This was a lot better than the high chimneys and the metal pipes of the continuously operating factories in the oil ports.

The noise barrier and the dense vegetation were more important to the buffer's function than its width; there were no more than 200 m between the satellite city and the factory. That proved tragic in 1968, when an explosion at Shell killed two employees and shattered thousands of windows in Hoogvliet and its surroundings. Over time, more and more infrastructure was squeezed into the buffer, in the form of a narrow passage for the hinterland transport of the western ports. A section was recently overhauled to make room for the expanded A15 and an underground pipe for waste heat.

Residential areas are no longer built so close to the petrochemical industrial zone, and there is no longer a need for personnel to commute to work by bicycle in great numbers. But it is also no longer necessary to hide the industry. A part of the buffer is now a post-modern park area and eroded paths show that Hoogvlieters enjoy climbing the noise barrier and the hill in the park, to admire the spectacle of the industrial environment.

– PPW –

Jannes Linders

EBB AND FLOW

North Sea, at Prinses Máximaweg, Maasvlakte 2.
First January storm of 2015, wind power 9. Many port operations were shut down.

Central Maasvlakte seen from E.ON Powerplant 3.
The smoking chimneys of the great industrial revolution, and the wind turbines of a possible post-industrial revolution.

Intersection of Ridderkerk (top right) and Reeweg, A15, Distripark Eemhaven.
Two of the most beautiful intersections: many decibels and quiet nature as a huge sandwich.

e van wijk
UNIT45
samskip

 Europaweg, entrance area to the ECT Delta Terminal.

MAERSK

Maasvlakte 2 and the North Sea.
The light here breaks in an air layer of ice crystals.

Maasvlakte 2.
Cranes under construction.

Malakkastraat, the bus transfer point, at 06:30 a.m.
Here, the changes in a continuous shift system are synchronized.

Vondelingenweg, petrochemical industry at Botlek.
The roaring, fire-breathing dragon along the A15.

Prinsessenhavens, Euromax Terminal.
The falconer.

The Oeverbos, between Vlaardingen and Maassluis.
Cars and scooters can travel to the edge of the New Waterway, and beyond.

North Sea at Prinses Máximaweg, Maasvlakte 2.
Forty-eight hours after the big storm in July of 2015, the boats sail back inside.

ESTIMATED TIME OF ARRIVAL

Port Development 1940 to Present

Marinke Steenhuis

CATASTROPHE AND RECOVERY

The city centre of Rotterdam was heavily damaged by German bombs in 1940. Four years later, with the Wehrmacht on the retreat, they left 7 km of blown-up quayside behind, a third of the total length of the embankments. Ships were scuttled to block the port entrance. The energy of reconstruction brought with it a great upturn in the pace of investment and modernization. The port was prioritized in the rebuilding of the city. Quite apart from the wrecked quays, nearly half of the warehouses had been destroyed. The four municipal docks had been deliberately sunk and dozens of cranes were standing idle on the banks. A specially-formed Port Restoration Committee, chaired by K.P. van der Mandele, who had become legendary as President of the Chamber of Commerce, set out the lines for the reconstruction and modernization of the port. The pre-war investment in the Waalhaven now paid dividends, especially once the Maas Tunnel was opened in 1942. Now, with the connection in place, new stevedore firms had the confidence to set up shop down there.

On 6 June 1945, just a month after the final liberation of the Netherlands, Frans Swarttouw was back in the port launching his first post-war ship, one commissioned by the Dutch government. His stevedore firm on Waalhaven Pier 6 had suffered great wartime damage, but was speedily back on its feet. This celebrated transhipment company profited from the ongoing mechanization of the loading and unloading of bulk cargoes. Yet this was a short-lived boom, because a revolutionary new development was just around the corner: the

01

shipping container, whose invention would turn the whole port economy on its head, sweeping away old companies and offering new ones an opportunity. That is always the story of the Port of Rotterdam: constant change as new technologies make their mark. After 1945, for example, the country did away with the old barge shuttles between Rotterdam and The Hague and Delft: freight would now be sent much more quickly in heavy goods vehicles, which were also able to reach the destination more accurately. The street names of Haagseveer and Delftseveer remain in Rotterdam, harking back to these old ferries.

Reconstruction of the port was finished as soon as 1949. It was efficiently organized by a joint venture between five Dutch and three British contractor firms specializing in concrete and steel constructions, trading under the name of

01

When it opened in October 1953, the Lijnbaan was the first shopping promenade in the Netherlands, and was actually the first car-free shopping promenade in the world. The name comes from the rope-making company that was based here between 1667 and 1845.

Maatschappij Havenherstel N.V. (Port Reconstruction Company Ltd.). The ravaged city centre was also rebuilt, but it was not the Rotterdam of before the Luftwaffe. The wide, car-centred boulevards such as Blaak and Leuvehaven, the traffic squares such as Hofplein and the new promenades of boutiques such as the Lijnbaan turned the city into a piece of Americana in which a new and un-Dutch lifestyle was born. Engineer Jan Hoogstad, the architect of the new plan and a native son of Rotterdam himself, acknowledged in his cityscape the influence of the port, which with its docks, sluices and ships had predetermined the scale at which the city should be rebuilt.

THE SILICON VALLEY OF TRANSHIPMENT AND OIL: 1945-1966

Oil drums, more oil drums and chemical plants with a maze of tubes sticking out of them; ports full of tankers and white warehouses. Nothing in Botlek or Europoort nowadays survives to show the agricultural history of Rozenburg and Welplaat, the two river islands in the Maas that were transformed into immense transhipment and industrial harbours in the 1950s. While tankers averaged around 16,000 tonnes in 1945, just ten years later there were already 30,000-tonners in service. With this in mind, one is bowled over by the prescience that Koomans had as Director of the Port Authority when he had the Botlek area constructed to take future tankers more than twice that girth: ships of up to 65,000 tonnes with a 12-m draft. Koomans managed to tempt no fewer than five refineries to relocate to the new harbours. The first started production as early as 1936: the refinery commonly known as Shell Pernis, or more formally the Koninklijke Nederlandsche Petroleum Maatschappij (Royal Dutch Petroleum Company). The refining of petrol was already proving hugely lucrative, as it was the fuel of the next great commodity, the automobile. One of the key economic pillars of the modern Port of Rotterdam, the petrochemical industry, had been erected. The motion that Koomans drafted in 1947 described this industry as the 'firm basis' for the supply of loading capacity, such that the purely transhipment-based model of old was now being extended to a model involving factories situated in the port, whose manufacture of products would provide a supplementary kind of employment. This would boost the job chances of the population in the Rhine estuary.

FIRST MAMMOTH TANKER

Every aspect of the port is related to geopolitics. After Prime Minister Mossadeq of Persia (Iran) expropriated the Iranian Oil Corporation in 1951, the director of the Port Authority Koomans managed to get the Iranians interested in Rotterdam. Verolme also took advantage of this situation; at their shipyard in the Botlek, the world's first mammoth tanker was baptized the *Resah Shah the Great* in 1958. The phrase 'mammoth tanker', incidentally, was coined by Verolme's communications officer.

Koomans' vision determined the whole life course of Celi Wiersma-Barendregt, born in 1943 as one of five children of a farming family at Kastanje Hoeve in the Oude Polder. This polder was on the island of Rozenburg, whose farmland was in the 1950s largely sacrificed to be excavated for the Botlek industrial zone. As a 16-year-old, Celi began keeping a scrapbook of her world, her family, her village, her school, De Beer Nature Reserve, and press cuttings about her father Arie, an alderman for the area. This scrapbook documents the personally felt effects of the enormous growth in the port in the 1950s and 1960s. The port really was snapping at their heels: it gobbled up their farmhouse not once but twice, with the buildings demolished to make way for industry. The farmhouse of Kastanje Hoeve had stood just outside the village of Blankenburg on the eastern tip of the island since 1815. Celi's father, Arie Barendregt (1913-1973), who served the community as alderman, water board chairman and acting mayor, found himself in a time of real uncertainty. He had the honour of welcoming Queen Juliana to the damming works on the Brielse Maas in 1951, and also took charge of rescue and reconstruction efforts when the North Sea Flood hit in

02

02

From the age of 16, Celi Wiersma-Barendregt (second from the right) kept a scrapbook of her daily life. She grew up on the farm known as Kastanje Hoeve, near the village of Blankenburg.

03

De Beer was a paradise for birds; species such as the sandwich tern, the gull-billed tern, the ruff, the dotterel and the common tern came here in droves. Celi Barendregt devoted two pages of her scrapbook to these birds.

04

On 13 September 1958, the farmers of Blankenburg listen to the explanation of new plans for their country, in the presence of Queen Juliana (front row on the right, wearing a hat). After this speech, the Queen used a siren as a signal to officially mark the start of the dredger.

05

The farmers who were given a new farm in the Krimpolder in Rozenburg were only briefly able to work and live here. Only two years after the move, the expansion of the Europoort was announced.

03

ROTTERDAM
NIEUWE
WATERWEG
MAAS

04

pl 120
12.1
pl 121 (Hulppaal)
pl 60
1.9
De
Beer
pl 122 (Hulppaal)
pl 122 (Hulppaal)
pl 50
pl 122½ (Hulppaal)
Hp 122½
Hp 123
Hp 123½
pl 123 (Hulppaal)
pl 123½ (Hulppaal)
Gat
1.3
0.7
Marijs
Keur
Schelpen weg
De Pan of Krimpolder
1.0
A. Geluk
Krimweg
Vogelweg
Kooi
weg
Stroo
Francke
Poppe
pl 30
pl 20
Dieleman
Vogelplaat
v. Nieuwenhuijzen
Klutenweg
pl 1029
Hp 365
Buiten
pl 1028
Nieuwlandsche
Polder
De Bruijne
Watertoren
De Beer
R. Geluk
Scheur
v. Kempen
Scheurpolderweg
0.7
polder
WATERWEG
v. Strien
Afgedamde
Scheur
Scheurdam
Sl
4.3
Hp 360
Noordbankpolder
0.5
3.4
3.5
Jantjesplaat
0.3
polder
pl 10
Barendregt
Overbaakseweg
Zeehonden
BRIELSCHE

05

06

06

Cèlil Barendregt on a tour through the 1,300 ha De Beer Nature Reserve, c. 1956.

07

On 28 November 1957, the city council discusses the Europoort plan. It would include the construction of grounds for a blast furnace and a steel plant, and the storage and transhipment of coal, ores and petroleum. It would also house a repair company for large vessels. In addition, the necessary infrastructure was included in the plan, including a pipeline route and a lateral canal (the Hartelkanaal). The first outlines of the Maasvlakte are already visible in this plan.

08

Sketch by landscape architects from the office of the Staatsbosbeheer, who manage nature reserves in the Netherlands, of the effect of the 1953 North Sea Flood on the sea inlets of the Zeeland and South Holland coast. Here a variant in which all inlets are closed. The importance of the port was served by an open Westerschelde – and so it was done.

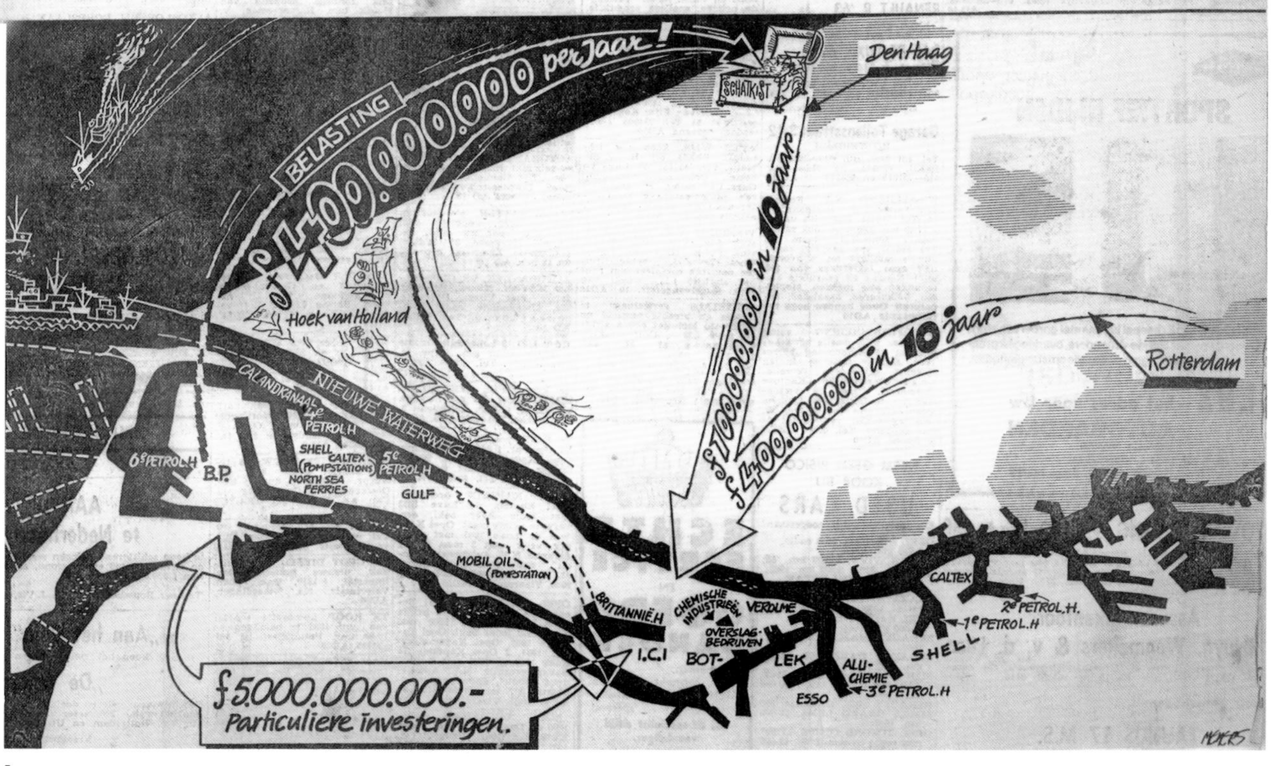

07

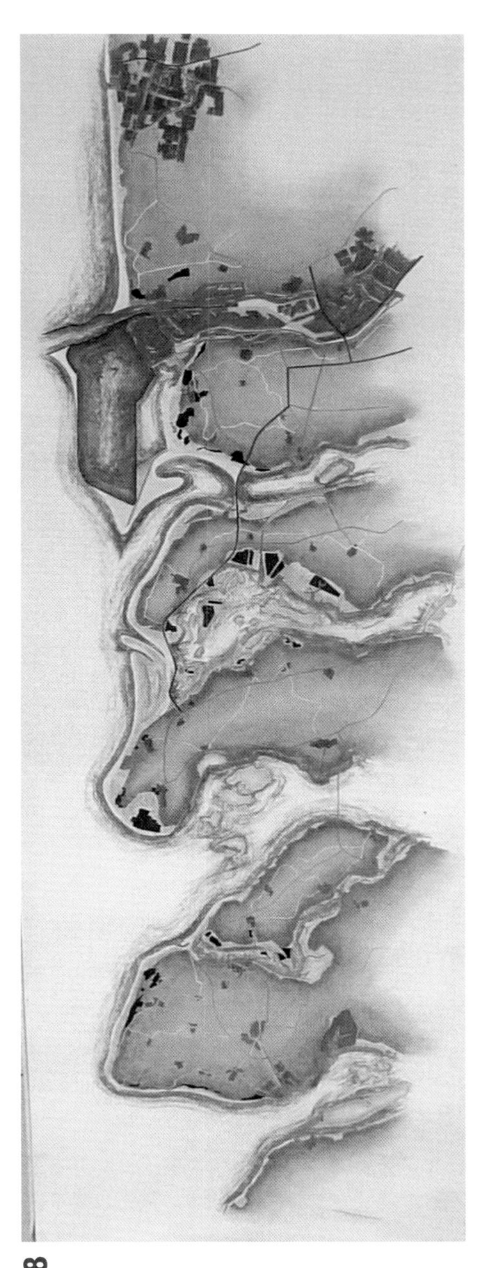

08

1953. Four years later, Arie found his beloved farmhouse scheduled for demolition together with the Oude Polder and the entire village of Blankenburg in order to let the industries of Botlek expand. The works doubled the area of allocable port industrial estate from 742 hectares to approximately 1,440 hectares. As construction of the Botlek Bridge encountered delays, it was in fact not until 1955 that the new industrial space came into use. During the years of delay, the plans continued to be amended: they grew in step with the expanding size of tankers during the period.

Arie was apportioned a completely new farm on Rozenburg, in the Krimpolder. In her scrapbook, Celi shows us photos of the birds in the adjacent De Beer Nature Reserve, over 1,000 hectares in size, where the family made trips. It would be less than two years before the Queen was back in the area, this time for the official opening on 13 September 1958 of the next phase of expansion. Laid out beside Botlek, this one was to be given the ambitious name of Europoort. The industrial plots in Botlek had proven to sell out within nine months, thanks to the runaway success of that new invention, the oil pipeline, which allowed huge quantities of the black stuff to be pumped straight off the tankers, deep into Germany. Actually, Posthuma, at the helm of the Port Authority, had secured the future of Rotterdam as a port by the skin of its teeth. He had responded with lightning speed to pip at the post West Germany's plans to build major tank installations of its own at Wilhelmshaven. In joint efforts with the Zuid-Holland Provincial Planning Service, he managed to lure the oil companies to Rotterdam rather than the north German coast. Time was of the essence, though.

EUROPOORT HUNGRY FOR LAND

Never mind that J.P. Thijsse had sung of the beauties of De Beer Nature Reserve in his Verkade Albums, and never mind either that the farmland here had a total annual yield of 1 million guilders; the new designated use for the area at the mouth of the New Waterway placed all this in the shade. 'Thirteen farmers, a gamekeeper with a patch of land, a duck decoy and twenty-five agricultural labourers, and their families one and all, have to evacuate West-Rozenburg because Rotterdam is determined to dig out new harbours and set up industrial estates', reported a newspaper article at the time. Among those 13 farmers, there were actually some who had only refounded their businesses here after the 1953 floods, and who thus had ended up working this land for just five years or so. 'Europoort has become a crock of gold. Big money is now dancing across this reclaimed land as friskily as the calves were leaping through the grass here until recently', a newspaper commented in 1966. The Rotterdam-Rhine oil pipeline into the hinterland, with its capacity of 20 million tonnes, had already been in use for some years by now. In all, industry invested 5 billion guilders in this project (by way of comparison, the total bill for the Delta Works, which were spread out over three decades beginning at that same time, was only 3 billion guilders). Huge housing estates for thousands of workers rose up in what had been the peripheral commuter towns of Hoogvliet and Spijkenisse. For farmers, there was no future on the island of Rozenburg now. The Barendregt family left for newly-reclaimed farmland in the Noordoostpolder. The family drama continues to our day: the very last farm on Rozenburg, where Celi's mother was born, is now threatened with compulsory purchase for the construction of the Blankenburg Tunnel.

AROUND THE CORNER FROM THE HARBOUR: LIVING IN THE MIDST OF PORT INDUSTRIES

The Port of Rotterdam surpassed New York in 1962 to become the world's biggest. The city's neighbourhoods grew concomitantly with the expansion of the port facilities: massive new housing estates were designed around Europoort and Botlek by the City Council's Urban Development and Regeneration Department. Spijkenisse, Hoogvliet and a new garden village on what was left of Rozenburg were designed as model living environments for the new workers drawn from across the Netherlands who would people them. Thus generations grew up with the port just around the corner, and with the striking landscape of night lights part of their identity. Across the river, in Pernis, Shell made special architectural efforts for its 1957 headquarters, designed by in-house architect Kees Abspoel (1899-1970).

Not everything went smoothly in this growth spurt of city and port alike, with households and industry situated side by side. The drinking water was one of the more crucial amenities that was not up to scratch. Bart Kuipers, now Director of Business Development and Business Relations at Erasmus Smart Port Rotterdam, grew up in Spijkenisse, where his father had settled as a family doctor for this new town on the southern edge of the conurbation. 'The tap water was undrinkable!' Kuipers recalls. 'Relatives from the north of the country used to come down to us with their boots full of jerry cans of water.' The water purification plant on Berenplaat in the Biesbosch marshes, opened in 1966, put an end to this third-world situation and ensured the whole Rhine estuary was supplied with clean potable water. The metro, too – Rotterdam's opened in 1968, making it the first in the Netherlands – was part of the modernization of the infrastructure in the residential parts of the city.

60

At the Berenplaat, water taken from the Biesbosch has been purified into drinking water since 1966 for the majority of South Holland's islands and the Rijnmond area. With its annual capacity of about 100 million m^3, the Berenplaat is currently the largest drinking water facility in the Netherlands.

ELEGANCE ON THE MAAS

In the 1950s, the focus of attention was on spatial quality: along Vlaarding's Maasboulevard, for example, where the modern Delta Hotel, designed by architect J.W. Boks (1904-1986), has stood in a park-like setting since 1955.

Locals remember an explosion at Shell Pernis as *de grote plof*, 'the big boom'. Across the south and west of the conurbation, in Hoogvliet, Vlaardingen, Spijkenisse, Pernis and Rotterdam-Zuid, residents were rudely awakened by the noise of this accident in the early morning of 20 January 1968. This being still the height of the Cold War, some thought the Soviets were invading. The resulting inferno was the biggest the city had seen since the Luftwaffe raid and appeared to be unquenchable. Two Shell men lost their lives and nearly 80 people received burns or were injured by flying glass. Tens of thousands of panes of glass, still gleaming in the newly-built apartments around the refinery, were gone in an instant. But the port still needed to grow, and Mayor Thomassen arranged large trade delegations to spread the fame of Rotterdam Europoort to the four corners of the world.

IN THE WAKE OF UPSIZING: THE SHIPPING CONTAINER ARRIVES, 1966

APM's fully-automated container terminal on Coloradoweg on the Second Maasvlakte is currently the most impressive facility in a development that began in 1966. Because channels have had to be dredged ever deeper to be navigable to the latest ships, this is where the largest vessels are now loaded and unloaded, those with the unimaginable capacity of 20,000 containers. The whole loading and unloading process is pre-programmed. It was 50 years ago that the

10

The 'car beach' in Oostvoorne was the only beach in the Netherlands that was accessible to cars. Even before the Second World War, it was common to drive cars on this beach. The elderly and the disabled, kite surfers and kite flyers all gathered here on the beach. The beach was closed in 2004.

11
Recreation at Noorderhoofd at Hook of Holland.

12

12

The inhabitants of Rotterdam live in the biggest port in the world. This ensures that this landscape forms part of their identity. Both residents and recreational visitors are attracted to this industrial spectacle.

13 – 14

Illustrations showing Rotterdammers taking possession of their new landscape. In this strange biotope, they still managed to find places for recreation.

13

14

A MAST, NOT TOWER

The first edition of the Floriade was held in 1960, in Rotterdam's Het Park. It was the first recognized global horticultural exhibition, and the Euromast was built for this occasion. The name 'mast' was coined by the director of the Port Authority. At 107 m, it was the tallest structure in the city at the time.

FIRST CONTAINERS

Containers made it possible to load and unload goods quickly: the *MS Fairland* moored in Rotterdam on 3 May 1966; in Bremen on 6 May, and in Scotland's Grangemouth on 8 May, where whiskey was loaded, destined for the American market.

first container ship was unloaded in this zone's Prinses Beatrixhaven: the *MS Fairland*, with just 226 containers on board. Containerization was a revolution in global goods transportation. It is a versatile unit, being equally at home on a ship, a train or a lorry. It signalled the end of backbreaking work for dockers and crane operators, who previously had had to clear out each ship's hold. Even more radically, it put an end to the very concept of quayside warehouses, because the container itself is the construction. One year after the *MS Fairland* put in to the new port, the Europe Container Terminal (ECT) at Eemhaven had sprung up. This was where the five major mixed-cargo companies were based, including Swarttouw Brothers, and Dutch Railways was also present. The arrival of container shipping brought about nothing less than the globalization of the economy, and it precipitated innovation and collaboration in the Port of Rotterdam. This was one of the few port cities in the world with enough depth of water to accommodate the draft of the ever larger container ships being built. To make loading and unloading the ships as quick and efficient as possible, and to give the oil and steel industry the chance to get ahead, another new port facility was needed. This became the First Maasvlakte, built out in the North Sea and opened in 1973 to be acclaimed as the port engineers' latest triumph of civil engineering and water management.

Plan for five power plants of 1,000 MW each, on the Maasvlakte. Ultimately only one was built.

The VU Working Group for the Wereld Natuur Fonds (the Dutch branch of the World Wildlife Fund) holds a protest march, and drops off bottles of polluted Rhine water at the German and French consulates. Amsterdam, the Netherlands, 28 June 1969.

ENTERING DIFFERENT WATERS: FROM CONSTRUCTION TO MANAGEMENT, 1970-1980

There are two power stations on Coloradoweg on the First Maasvlakte. The older of them was commissioned in 1973. Its machinery was built by the Rotterdam Dry Dock Company but the issue of its coolant water was solved far away in the Noordoostpolder, a reclaimed area of the old Zuiderzee. That was where the National Hydrological Laboratory (Waterloopkundig Laboratorium) had been established, a massive collection of open-air scale models of waterworks from around the world. The engineers at the Laboratory were the brains behind the Delta Works, investigating harbours and rivers, coastlines and dikes, flow velocities and the movements of coolant water. The scale models are screened off as far as possible from the effects of the wind by their location in a wood. The Rijks-waterstaat commissioned experiments at this polder lab to test every one of the complex port projects in development: the entry to the Botlekhaven, the Tweede Petroleumhaven, the Europoort breakwaters, and the issue of how best to circulate coolant water around the Maasvlakte power station.

Ton van der Meulen, who served the organization as project engineer from 1963 to 1968 and then became Head of Barrier Works through the key years of 1968 to 1986, is still fondly remembered by his old colleagues as 'Mr Waterweg', because in the former of those capacities he managed the 600-m-long scale model of the Waterweg canal in Rotterdam. As Van der Meulen recalls:

'They were saying we needed a 5,000-megawatt power supply on the Maasvlakte. That would entail five power stations each with a capacity of 1,000 MW. The reason for this was that a second blast furnace was expected to go up. A circuit of power stations that big is built in phases, each containing a pair of 500 MW facilities with their own coolant circulation. The coolant obviously needs to be fed in somewhere and has to be able to run back into the sea, and that operation would necessitate cutting a channel through the ring dike around the Maasvlakte. Water safety considerations meant there was only one location where that could be done, so we came up with a design of a single cooling pond

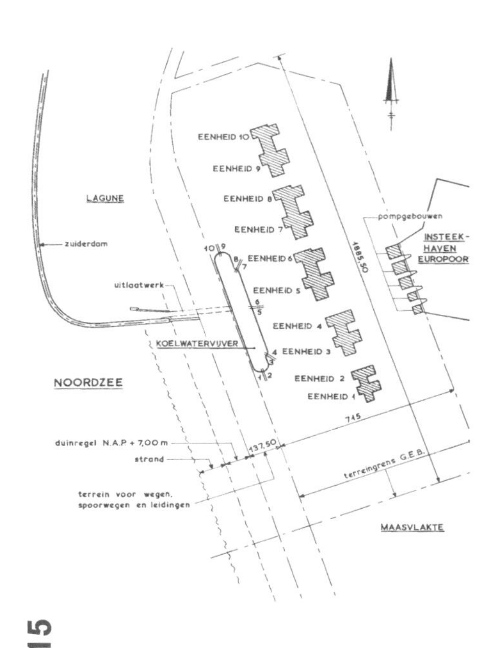

15

fed by five intakes and a single outlet to the North Sea. We investigated how the flow was going to work by scattering paper shreds on the water and taking long-exposure photographs. The culture at our lab was, if you had a good idea, you were free to put it into practice. In the end, only one power station needed to be built. I call it the 1970s craze: we were so thirsty for energy back then'.

THE PORT LANDSCAPE SET IN STONE THROUGH LEGISLATION

The growth of the Dutch welfare state and the construction of new housing for millions of the country's people gave force to the plea for new planning procedures, ones that were more in line with the varying needs that there were. The Rhine estuary, under its Dutch name of Rijnmond, became a national byword for both the wonders and the terrors of urbanized society, with people corralled into zones around their places of work. Living in the thick of the petrochemical industry proved none too pleasant for those who had to. A 1965 planning law, the Wet op de Ruimtelijke Ordening (Spatial planning act), opened up a new world of possibilities for urban planners. The following year, Gulf's refinery had a temporary closure order slapped on it as a result of the nuisance caused by noxious gas emissions at Botlek and Europoort. The planning principle of 'concentrated deconcentration' was followed, and a 1966 follow-up white paper on planning issues heralded the nationwide concept of the *groeikern* (growth hub), a new way of coping with the relentless demographic growth by designating certain locations to become modern garden cities where folk could happily both live and work surrounded by greenery. The aim of this was to spread the job opportunities out from the Randstad to the north, south and east of the country, and accordingly no further expansion was envisaged for the Port of Rotterdam. The Port Authority was up in arms about this white paper. Mayor Thomassen called it 'a biased piece of work that can best be filed in the National Archives as quickly as possible'. Dutch society was inexorably becoming permeated with a new sense of the good life: one in which it was more essential than before to weigh competing interests in society against each other. With government grants, many young people were becoming the first in their families to go off and study, and some of this newly academic generation became taken with the notion that unbridled growth had reached saturation point. The Club of Rome's

16

17

THE GATDAM IN BRIELLE

The Gatdam in Brielle, built in 1966, meant that the connection with the North Sea was severed. This created the Oostvoornse Meer, which was intended as a nature reserve. The Brielse Gat became a sea inlet area, a tidal area with wet and dry sections.

Landscape architects were involved in the design of the port. This is a sketch for the design of the intertidal area of the Brielse Gat.

1972 report *The Limits to Growth* drew on its members' prominence in the world of science to convince the public that man's pattern of consumption of fossil fuels would ultimately exert a depletive effect on the global ecosystem.

THE FIRST MAASVLAKTE

The First Maasvlakte had been on the cards for years already before Rijkswaterstaat built the first dams for it in 1965. J. van Veen (1893-1959), a doctor of engineering and the 'Father of the Delta Works', was the first in his profession to propose that the Dutch coast be pushed back. By a coincidence of history, he finished his plan for the closing-off of the sea inlets in the Rhine estuary just two days before the disastrous North Sea Flood of 1 February 1953. He was prompted not only by considerations of safety but also by how to counteract the effects of silting on local farming: the brackish water table was rendering the fields infertile. When, in the mid-1950s, he proposed construction of the Maasvlakte to extend the port, his civil service employer, Rijkswaterstaat, found the suggestion too radical. Consequently, the Europoort project was cobbled together instead in double quick time. It could be left until later to haggle over the form and scope of the port extension, but it was quite obvious that Rotterdam was going to obtain a seawards add-on.

However, even though the island farmers of Rozenburg went without a fuss to their new land to make way for Botlek and Europoort, and although the 1,300-hectare De Beer Nature Reserve was quietly sacrificed to the harbour, the planning and construction hit a squall anyway. The nature protection lobby, which was really coming into its own in these years, was outraged at the likely fate of the Brielse Gat, the eastern end of the stretch of water known as the Brielse Maas. It had already in fact become a bay since the damming undertaken for Europoort in 1951. The loss of this valuable integral tidal range was something the nature movement could not countenance. Having sought advice from the Government Service for the Implementation of the National Plan (Rijksdienst voor het Nationale Plan), the government settled on the option of installing a 'line of demarcation' between the areas given over to port activities and the zones left for nature to take its course. This dividing line ran from the northern part of the Brielse Dam through to the Westplaat. This, in 1964, was the moment at which the earlier post-war policy, one which had gone all out for growth, shifted emphasis to one that paid more attention to other interests besides. One of the consequences of this shift was that rows of sand dunes were put up alongside industrial estates on the Maasvlakte, for instance north of the Maasvlakte Oil Terminal (the country's strategic oil reserve). The imposition of the line of demarcation heralded a new era in the history of the port landscape. From now on, a nature-inclusive approach had become the basic premise in planning policy, and this brought a wonderful mix of ambiances and qualities to the area.

FOSSILS

The First Maasvlakte was constructed by laying a ring dike: an enclosure into which North Sea sand was poured. This sand, which has since become the foundation of the Maasvlakte, contains many fossils. One of the interesting aspects of it is that due to the sequence of construction, with the lowest extracted sand ending up on top, what was once the oldest layer of all in the North Sea (when it was in fact not yet a sea but a swampy plain) is now at sea level for visitors to interact with. The blast furnace and steel industry that

Koomans dreamed of back in 1957 as director of the Port Authority never arrived: it turned out to be too much of a challenge to identify in the Rijnmond area the large number of specialist workers needed, and local inhabitants were uneasy about the risks to the environment and their health and safety. As it transpired, the 3,000 hectares of the First Maasvlakte were in great demand by the chemical industry, including petrochemicals, and also sought-after for the

Plan 2000+, presented by the Port Authority director Ir. F. Posthuma, symbolized the megalomania of the port managers. The oil crisis of 1973 and the changing social tide put an end to this vision.

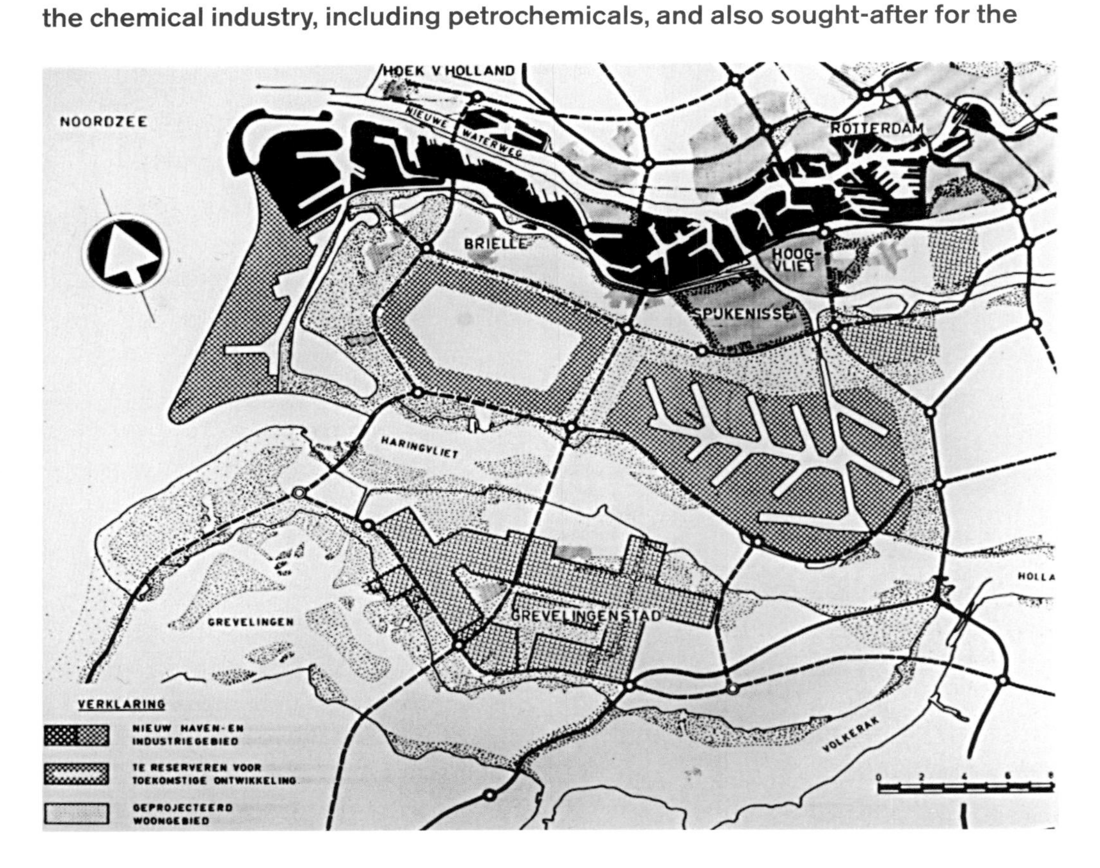

18

storage and transhipment of dry mixed cargoes. Shipping container company ECT built the Delta Terminal there. It was a project initially greeted with scepticism, but the large-scale approach of this terminal gave Rotterdam a competitive edge. In the best Dutch tradition of multipurpose design and of allowing various claims on space to coexist, the Maasvlakte was built replete with depots to store polluted river and harbour silt. These are named the Slufter (Tidal Inlet) and the Papegaaienbek (Parrot's Beak).

Engineer F. Posthuma (1913-1986), who took over the directorship of the Port Authority after Koomans's retirement in 1959 and who continued coordinating the latter's large-scale port expansions by persuading one company after another to relocate, did not last long before he came to be regarded as the embodiment of all that was untoward about the Port of Rotterdam. The unveiling of his *Plan 2000+* in 1969 proved to be his undoing. Presented to the public as a discussion paper for 'a harmonious development of port and industry facilities and of recreational and residential areas in the northern part of the Delta', a closer look at the map revealed that the idea behind this vision for 2000 and beyond was to dig out the whole island of Voorne-Putten and to turn the Hoeksche Waard, famous for its trees, into an industrial complex and airport. Public opinion wrote Posthuma off as a relic from a past era; the port directors' fixation on further expansion was derided as a delusion of grandeur. Then the 1973 oil crisis struck; everything ground to a halt for a while. For just a temporary breather, the city had not a shortage of space but a surplus of it. Just before the crisis hit, Cornelis Verolme had had one of the world's biggest dry docks constructed on Professor Gerbrandyweg in Botlek, expecting that tankers and other ships would soon be built in it. Yet the spike in the oil price had turned new shipbuilding into a loss leader. It had all become too much for Posthuma: on the day that the Arabs raised their oil price, he handed in his notice as Port Authority director.

KUNST EN VAARWERK

In 1979, the artists collective Kunst en Vaarwerk began to enrich the port landscape with monumental objects that often had an alienating or humorous effect. 'We wanted to use art to make the identity of the city more visible. Everything was grey, and gradually we added a bit of life to it,' said artist Cor Kraat in a filmed portrait from 1980, on YouTube.

CONSOLIDATION AND A NEW-FOUND FLAIR

After the dizzying growth of the 1950s and 1960s, it seemed for a while that the port's major expansion needs were sorted with the completion of the First Maasvlakte. This was a welcome pause for reflection on the landscape that the port had been left with after the last draglines drove off. How did the area now look in reality: Had the planners really paid sufficient attention to safety and environmental concerns? It was now high time that the port's heritage be looked at in fresh ways, and accordingly new roads and tracks were built through the area and a tree-planting plan was implemented. The area around the old mediaeval port was not left out, either: it was reimagined with patios, residential new-build and the above-street constructions over the open space of Blaak. Engineer H. Molenaar, who directed the Port Authority from 1980 to 1992, recognized that enhancement of quality and 'added value' was called for on all fronts: the current business model, heavy on transhipment, oil and refineries, had to be diversified. What he grasped was that Rotterdam's distinctive edge no longer needed to depend on the hardware of its port. However, physical infrastructure, docks, cranes and fairways are of course as necessary as ever they were, and this need was met in the form of the Botlek Tunnel, the modern container facilities on the Maasvlakte and three warehouse zones known as *distriparks*. Be that as it may, he realized, the city's added value now needed to spring from how it could uniquely leverage communications in the shipping industry. The port economy would still have a bright future awaiting it if it managed to combine processing information streams with the amassing of social networks. These insights were gold dust, coming as they did during the infancy of the Internet. Nowadays, the competitiveness of Rotterdam as a port does indeed largely come from the 'software' of human interaction, which, enabled by technology, is becoming ever quicker and smarter.

MONUMENTAL WINDSHIELD

Municipal architect Maarten Struijs designed the monumental windshield along the Caland Canal, together with artist Frans de Wit. The structure, completed in 1985, was meant to ensure that the canal remained navigable even at a wind speed greater than 5 BF.

A second area of focus was image management, since this was increasingly tarnished as far as the Dutch public was concerned. The nation had come to regard the old qualities of the Port of Rotterdam as having vanished from view, both literally and figuratively. Mergers and takeovers of classic port companies had become run of the mill, and locating your business in the industrial districts of Botlek or Europoort had come to be seen as the kiss of death. For Rotterdam City Council, which at this stage was still managing the Port Authority as a municipal service, it was becoming a matter of urgency that the world hear a new narrative about the port and city. It was doing little good that Rotterdam was the greatest port in the world while the city behind it lay dishevelled and neglected.

19

The 'Map of Riek' from 1987, depicting 30 urban and regional projects.

THE NEW ROTTERDAM

It was precisely at this low ebb about 30 years ago that the city managed to turn its fortunes around, and we are now reaping the benefits of those efforts. The groundwork was laid in the mid-1980s, when the city of Rotterdam was being run by an unusually fortuitous combination of good local government officials, highly-placed civil servants and urban planners. Led by Mayor Bram Peper, the executive of the City Council launched a new narrative for Rotterdam in 1982, one that linked the no-frills history of the port landscape to the future that the city was aiming for: that of offering a lively and unique investment climate. The city, still reeling from the abiding trauma of the 1940 blitzkrieg and from the incessant transformations of the port, now managed to repackage its experience from baggage to a unique selling point. The motto chosen was 'Het nieuwe Rotterdam', and for this new city a new Director of Urban Development was called for. The woman to fit the bill was Riek Bakker, who from 1985 on brought the worlds of city and port into dialogue with each other. The council hired a good number of designers and made a proper strategy of its land use policy. The great developments of Kop van Zuid, Hotel New York and the Erasmus Bridge are the fruits of this determination. Bakker well understood the qualities of the old inner-city docklands and their potential to bring about urban renewal. Rotterdam now began transforming itself from a city still encumbered with its reconstruction efforts, and hampered by its tendency to regard the port as the exclusive domain of engineers, into a city full of promise, creativity and verve, but equally a city with a welcoming and sociable nature.

DELTA WORKS REVISITED

The successes of the early 1990s have since been joined by a long list of more recent initiatives. Stadshavens, an urban port project to bring economic renewal to the campus of RDM and to M4H (Merwe-Vierhavens), would never have come about without this change of direction. The past 20 years have seen the port landscape continue to be subject to large-scale new projects. The Maeslantkering was opened in 1997, a barrier to protect the densely-populated shorelines of the New Waterway from high tides. It was designed as a movable barrier to allow free passage to shipping in normal times but to act as an emergency lock when the tides grew dangerously high. Its two enormous hollow doors flood with water when closed, turning them into giant plugs. When the threat has subsided, the doors are emptied again into the sea, allowing them to swing open outwards.

Another major works project in the same period was the Betuwe Line for rail freight: shipping container haulage on motorways had begun to encounter snags in the early 1990s, and the environmental toll of road freight was increasingly in

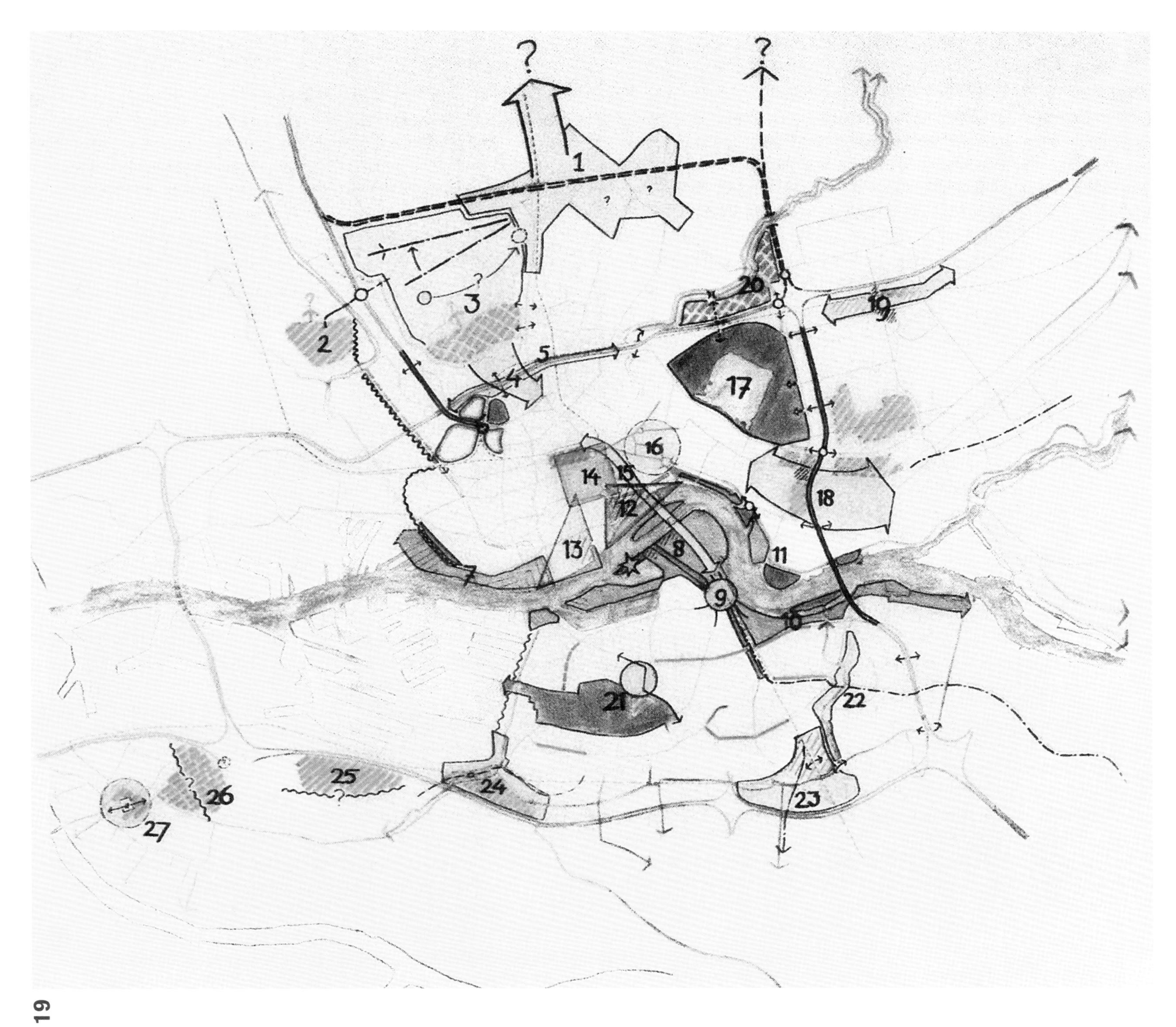

19

the public eye. Preparations thus got under way from 1995 onwards for the Betuwe Line, a 160-km goods railway to connect the Maasvlakte to the German border. Its 4.7 billion-euro cost made it far and away the largest and costliest infrastructure project that the Dutch government had ever undertaken. The original plan was that trains of shipping containers could pass straight through to the berths, but after the line was commissioned in 2007, it took a further eight years for the German network to build a rail spur into it. ECT (Europe Container Terminals), a deep-rooted and very innovative organization, is the largest customer of the Rotterdam Port Authority. A few years ago, it invested in inland European container terminals to allow shipping containers to be brought closer to the heart of the continent.

20

The *Kalmar*, a mid-sized Maersk container ship. Via Facebook, the whole world can follow the voyages of Captain Geert Sassen.

21

Spread over the Maasvlakte, West 8 designed a series of buildings for energy companies with various forms and colours. To make a cohesive design, a flat, glazed stone was chosen for the façades that reflects the Dutch sky and landscape. The buildings are part of an overall plan by West 8 for the visual presentation of the port.

KEEPING THE PORT LANDSCAPE THRIVING

Geert Sassen captains the *Kalmar*, a mid-sized cargo ship operated by Maersk. At 300 m long, she has a capacity of 7,000 containers, a 90,000-horsepower diesel engine, and an 8-m-diameter propeller. As Geert puts it: 'Compared with the 20,000-container ships, we're small fry, but if you compare us with the kind of ships that my father sailed in, or even the ones I began on in 1980, the *Kalmar* is still a monster vessel and a whole city in itself.' Her worldwide voyages can be followed on Facebook. Geert and his family live in Thailand, and Rotterdam is his working base. A couple of days before the 20-strong crew is due to set sail, they converge on the city from all corners of the world. They are from a great range of nationalities. The *Kalmar* plies a regular route between north-western Europe and India via the Red Sea and Dubai in the Persian Gulf. The round trip takes eight weeks. She has eight sister vessels, which dock at Rotterdam at very precisely 11:30 a.m. every Friday and depart again at 3 p.m. on Saturday. In other words, each Friday, the port reserves a 'window' of quayside space for one of these 300-m vessels. The ports of call along the *Kalmar*'s route also know down to the minute when to expect her in dock. There is hardly anything left nowadays of the individual character of ports and their port cultures. Everything is footloose and globalized, from the cargo loading and the crew recruitment to the traffic code for shipping.

But while globalization has continued apace in world trade, the Port of Rotterdam has actually become more welcoming. Here, on the Maasvlakte 2, is where the city loads and unloads its visiting ships, amid beaches, sand dunes and kite surfers. It is unique in the world that port industry is so cheek-by-jowl with recreation. Achieving a port landscape with added value for the public was very much what Hans Smits was looking for. He led the Port Authority from 2005 to 2014, just after its privatization in 2004. Smits says: 'I was in the Delta Service from 1975 to 1985, working on the Eastern Scheldt barrier project. That's the organizational culture that became my rule of thumb for the rest of my career. The Delta Service was an open, creative organization, not afraid to explore new ways of doing things. This made us hugely able to innovate and we managed back then to make a design reality of environmental care.'

This attitude was the determining factor as the Maasvlakte 2 was being built. After many years of prior studies, it ended up taking just three years to construct. The globalization of international shipping has prompted the Port Authority, as owner of the port facilities in Rotterdam, to start investing in the spatial quality of the port landscape. This was not a foregone conclusion by any

CLEANEST SEAWATER

Several Maersk container ships bring in the cleanest seawater from the Irish Sea, especially for the Rotterdam Zoo at Blijdorp. Here the water is pumped into the *Haaibaai*, a special water boat for the zoo. The *Haaibaai* then takes the water back to the zoo.

20

21

22

A large open sculpture represents the culmination of the construction of the Maasvlakte 2, overlooking the coast of Zeeland, the North Sea and the automated ports and dykes of the sea inlet. The concrete skeletons depict three stages of the dune formation, and continually take on new shapes as the sand shifts. It was inaugurated in July 2015.

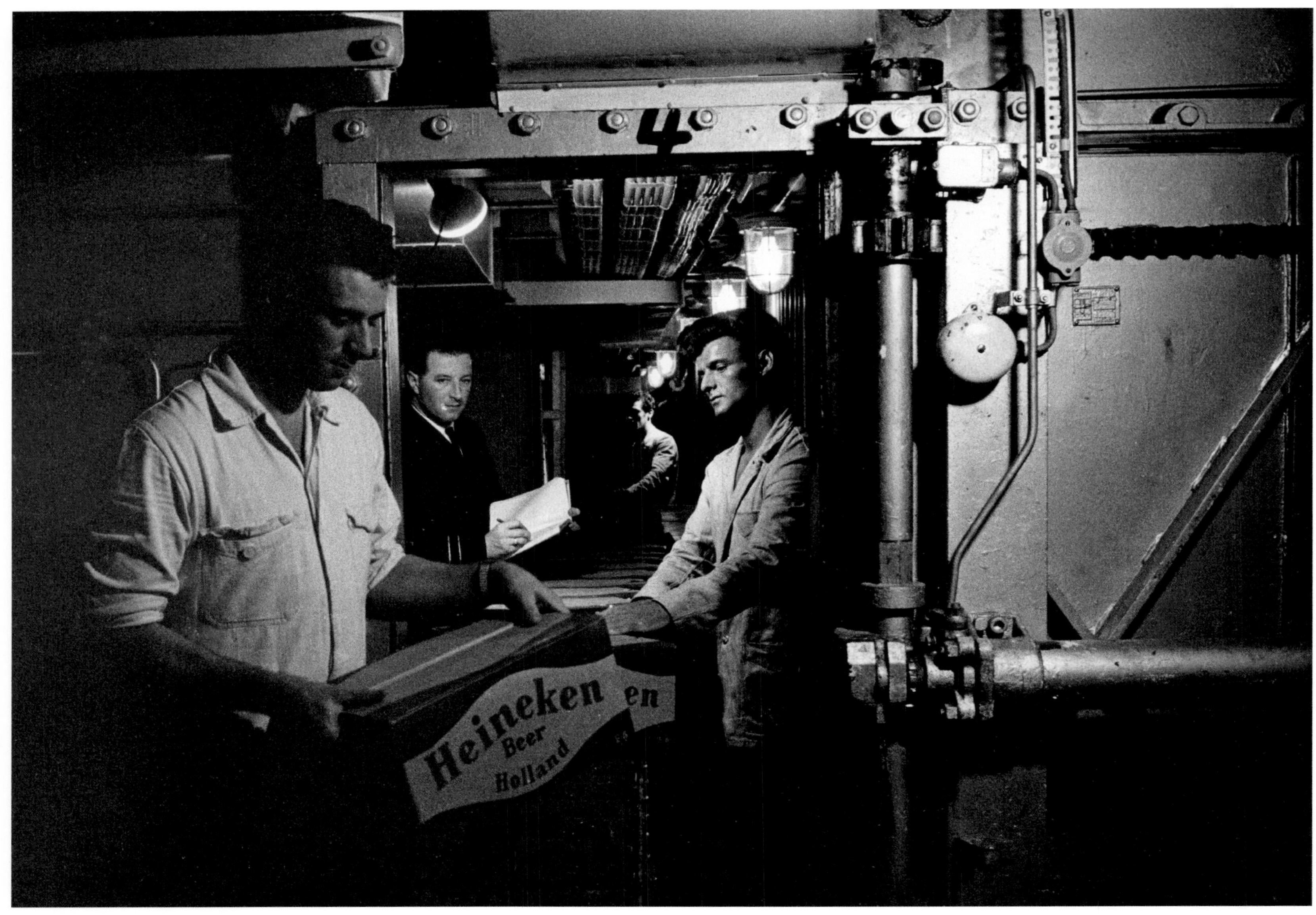

BUS IN THE MAAS

The amphibious Splashtours bus provides unique tours through Rotterdam. During a city tour, the most beautiful spots of Rotterdam are shown, after which the bus transforms into a boat, driving into the Maas.

23
Heineken is the second-largest customer of the port, after the oil companies. By 2020, the company will have shipped 50,000 containers of export beer to Rotterdam, from the brewery in Zoeterwoude and then via the transfer point in Alphen aan den Rijn.

means: for decades, Rotterdam was able to rest on its laurels as the world's biggest port. It was above all the rise of the Chinese ports and construction there that challenged this dominance. Much of the stuff of the daily news can be directly traced to issues at the port. The effects of fluctuations in oil prices, the challenges of climate change, the arrival of 3D printing (which is affecting container shipping): these are all developments that are continually causing change in the port landscape. Thanks to its success in weaving together the various claims that there are on space at the port, the Port of Rotterdam is increasingly a destination for all manner of people. The automation of many of the port's processes has left this a safer port landscape, and thus a more attractive one to other kinds of visitors, ones with no directly port-related purpose to attend here. It is forecast that in 20 years' time, we will be seeing 10,000 tourists head for the Maasvlakte on days when the weather is fine. This is a highly promising expectation and is already being fulfilled in some ways: Futureland, the Maasvlakte 2 information centre, can hardly keep up with the demand for children's parties. There is a website, Oervondstchecker,

UNCONSIDERED QUALITY

Jan Benthem, a member of the quality team for the port landscape since 2009: 'In the 1980s, I was flying in my plane to the Maasvlakte, and this barren plain was incredibly beautiful. I'm fascinated by industrial landscapes. What can you do there in terms of image quality? As an architect, you have to keep well away from that. Rationality and pragmatism, that's what matters. That and monitoring the non-quality, the unconsidered quality.'

where visitors can give notification of their fossil finds on the Maasvlakte beaches. What was originally conceived of as just a big fence has now been given shape as a new row of sand dunes to mark the transition from coastal zone to port facilities. Specially-designed dune viaducts allow both truck drivers on a day at work and visitors on a day off work to enjoy this mixed-use landscape. The port is criss-crossed by a 120-km network of cycle tracks. Restaurant Aan Zee on Lake Oost-Voorne gives patrons a dazzling view of the container terminal lights. The Slufter serves both as a repository for silt and as a waterfowl breeding ground. The embankments of the coastal road along the new shoreline are silos for spent batteries. Beereiland Nature Reserve doubles up as a shock absorber to enhance the safety of passage of the tankers laden with liquid natural gas. Windmills of various vintages are dotted here and there around the port area. It is a landscape for everyone: the A15 motorway slices through a spatial pattern that weaves together distribution policy, storage requirements, environmental reports and safety zones into a single superstructure with a logic and beauty all its own. This superstructure is boundless and scalable, and keeps generating new spatial projects as it goes. Nothing ever remains the same as it was; only that much is for sure.

Adriaan Geuze on the Port of Rotterdam

Marinke Steenhuis

PRIDE, COMFORT AND COMPASSION

My work and my life would be unimaginable without the port. The port formed my entire childhood. My father worked for Deutz and my mother comes from a family of inland waterway carriers, who moored their ship at the Dokhaven. My father took us to see the reclamation work that was done for the Europoort, which was already an attraction back then. I saw asphalt being poured on rockfill and took test rides on ships that sailed out to the oil rigs. A kind of pendulum test was performed on these ships, to test their stability.

I came to live in Rotterdam in about 1985, and used the Maasvlakte as my beach, and the seawall as my evening escape. The power station was there, along with the coal and ore terminals, the ultra-light aircraft, dog training with huskies, fossil hunters, the nude beach, a special beach for cars, the Rotterdammers fishing with flashlights, and the lorry drivers. There is so much space; the port is a planned place for all sorts of activities that would be unsuitable in the civilian city. It's a kind of orphanage for peripheral activities. I liked that. Seeing all those subcultures using the space helped me think about my profession in the anthropological sense, and about the phenomenon of the city. We Rotterdammers imbue the port with meaning. Its non-romantic character, the lack of scale; what does that mean for people? You are in the landscape, and you see things from your own perspective: you can see birds, a road sign, a factory – and everything seems to be on the same scale, which is an almost cosmic experience. At the same time it's an encyclopaedia that you can get a grip on, the story of the Pleistocene sands of the North Sea; the depth of time is readable, you can find a fossil and see a seal at the same time. The port area is the wilderness by the city, and forms an indispensable part of Rotterdam's identity. The yearning for the port is fundamental – Rotterdammers cannot yearn for cathedrals, because unfortunately things didn't turn out that way historically. That's why we have the cathedral of

Rotterdam in the form of Team CS at Rotterdam Central train station: the port, like a slow movie, is brought to life on a huge screen in the new station's main hall. The port is the soul of the city.

COLLATERAL BENEFITS

I grew up with the Spatial Planning Act. The reconstruction period was truly incredible; they were really making something there. But that feeing started to erode in the 1970s. The planning became a kind of paper planning, with structural visions and above all lots of procedures. To spell it out more concretely: we have 22 weak links in this country, places where the threat of flooding is acute. The amount of 70 million euros has already been spent on just one of those links, namely at Petten. That must mean something strange is going on.

In the early 1980s we made an excursion that took us to the port, to Schiphol airport, and to the IJsselmeer polders, places where the culture of making things was still very much alive. The Delta Service was already being dismantled at that point. But on that trip, it felt as if you were almost in the bastions of the ideal planning world. Of course it is no coincidence that the legacy in these places remained alive: the port, Schiphol, the polders – these are all products of pragmatic thinking, created by economic necessity. Innovations are the order of the day; without innovating, you'll fall behind very quickly. But something new was coming about: an attitude of making work out of work, a certain way of thinking that was always based on the question of what physical intervention can deliver in terms of bycatch. The construction of the Botlek, for example, also created the Brielse Maas recreation area, and the construction of the Maasvlakte created the Voornse Meer. The idea was not to think in terms of *collateral damage,* but rather *collateral benefits.* Could this spillage also deliver benefits for someone else? The Port of Rotterdam Authority has always been very efficient in this regard: its core objectives are focused on the future, on planning and implementation, on the business world and doing business, and on security. For me, the result is an enchanting beauty. If the port were to abandon this tradition of planning for a goal, then who else would do it? That is an existential issue for me. And as regards the slogan about citizens themselves having to do more: no, it's more that we have to do something with each other, the government and the citizens together. Leaving planning issues up to only the citizens is not something that we can afford in terms of population density. The Netherlands is the Gaza Strip: if everyone were to go their own way, then you would need a lot more space, on the scale of the United States.

HYPNOTIC CADENCE

For me, the port led to a new design paradigm, one that stays with me today. Insane and surreal places have their own beauty and enchantment. For the average Dutch person, the port is terra incognita, a place where another language is spoken, and the logic of the everyday world is far away. The entire Europoort was made on a drawing board with jumps of 15, 30 and 45 degrees. In the distant future, archaeologists will examine it like a kind of Stonehenge and develop all kinds of theories about it. For me, the power of ships and goods being transported form a hypnotic cadence. The fact that at the port lights by the Hook of Holland (at the tip of the Netherlands where the

New Waterway cuts into the country), you can use an app to track the location of ships, their names, how many tons they're carrying, where they come from – the mere fact that this goes on, 24 hours a day. Or that starting in Breda, a new lane appears along the A16 at every exit; it looks like the invasion of Normandy. Wherever you look, you see the wheel-wells of the lorries. Or that in the Botlek, all those steaming chemicals are clumped together in such close proximity; how is that even possible? The smell of the brackish water, the sediments, the ebb and flow, the colour of the water, but also the materials that you see lying everywhere: coal, ore, sand, gravel, the ancient materials that the earth is made of, all just lying there, for free.

One of my first designs, for the Schouwburgplein, can be traced directly to my nostalgia for the port, which offers a different kind of romance than the traditional urban decor. I used this particular beauty in the design, like a kind of poem dedicated to the port. Rotterdam needed it; it put the city on the map at a time when reconstruction had become a jaded concept, and the city was looking for a new self-awareness. At any rate, we didn't have much choice, because the garage under the square could not support any heavy material.

This turned out to be an aesthetic that had been missing from landscape architecture until that point, which was why the Museum of Modern Art in New York acquired the archives of this project. I feel akin to Yakov Tsjernikov (1889-1951), the architect of Russian constructivism. Everything that he drew was also scale-free, with a primitive form language, in which gravity does not seem to exist. That was the core of my recommendation to build Unilever's headquarters atop the former Blue Band dairy factory. When you remove gravity, everything becomes possible. The buildings then become characters, and this happens in the port as well: a bollard becomes an animal, and there are also giraffes and elephants. In his work, photographer Jannes Linders has always striven to make living forms from everyday places; in one of my favourite photos, you can see that very well.

The port for me is thus an orphanage and a storybook at once. The landscape provides a cinematic sensation that you will not experience anywhere else. At the same time, the port is an area where you encounter strange shapes, plus you have the Rotterdammers themselves roaming through it all. That evokes in me a great sense of melancholy, and also one of compassion; the comfort and pride that Rotterdam takes in the port. The power of its beauty, despite the ugliness: that is Rotterdam.

PORT PLACES

Frank de Kruif [FdK]
Isabelle Vries [IV]
Peter Paul Witsen [PPW]

photography
Jannes Linders
Siebe Swart

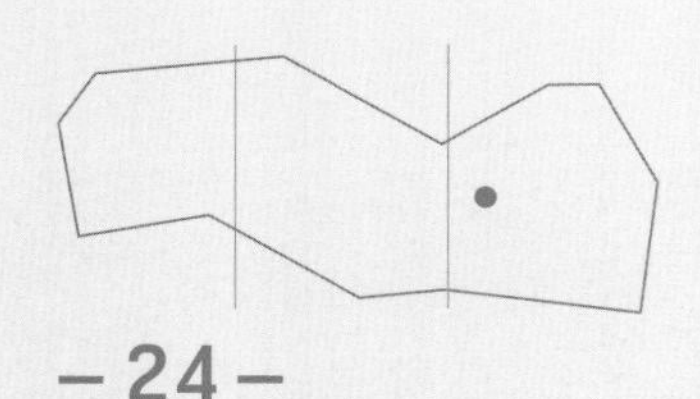

– 24 –

Marconistraat 80 \ Rotterdam

FRUITHAVEN

Transhipment of Refrigerated Cargo

Fruit is unloaded from a ship on the quay at the Rotterdam Fruit Wharf in the Merwehaven: this is a familiar sight in this part of the port, on the north bank, which is known as Rotterdam Fruit Port. Within a few years, the fruit handling here will be forced to disappear and move elsewhere, in order to make way for urban development.

As long as growers from the southern hemisphere have exported their crops to Europe and North America, this has been the way that fruit and other fresh produce (vegetables, fish, meat) have been shipped: in ships with holds in which the temperature is kept just above, or if necessary below, the freezing point. In these *reefers*, fruit is stacked in boxes and on pallets, and loading and unloading take place at specialized terminals.

Today, these refrigerated vessels, often painted white to reflect sunlight, can still be seen in the Merwehaven, often arriving from South Africa, Chile or other countries. But over the years their numbers have dropped significantly (which means that the quays can now be used as berths for inland shipping, as we see in the photo). In conventional fruit transhipment, the competition is fierce. For Rotterdam, it was major big blow when Antwerp snapped up the large-scale banana shipping in the 1970s. The Flemish port then took off as a leading fruit port, and has not been overtaken since.

But a later development was very much in Rotterdam's favour: the rise of the refrigerated container. Just as ordinary cargo was increasingly being transported in containers, this also began to happen with fresh goods. Vessels and terminals offer more and more connections for these reefer containers: in fact, they are no different than large refrigerators. As the market leader in container transhipment, Rotterdam naturally also became the largest handler of refrigerated containers.

The rise of the refrigerated container has had major consequences for how transhipment is done in the port. Much less than before, the conventional transhipment companies unload, store and distribute the fruit to the receivers, who are usually importers. For those fruit importers, it has become much easier to ship goods in a full container and to deliver the shipment themselves. It doesn't matter where the importer is located: in the refrigerated container, the cargo stays fresh. That is why inland shipping is increasingly taking over these containers from road transport. The only condition is that the inland barges and the hinterland terminals can provide electricity for the reefers.

Conventional fruit transhipment will not disappear. In the peak season, after harvesting, there is still a need to ship large quantities all at once in traditional refrigerator ships. The question is whether the market will continue to demand a central fruit port when the current one disappears. The Port Authority is expects the demand to still be there, and is trying to build a new complex on the south bank under the name Rotterdam Cool Port.

– FdK –

LETABA

LETABA

IMO 9016682
seaco
NO FUMAR

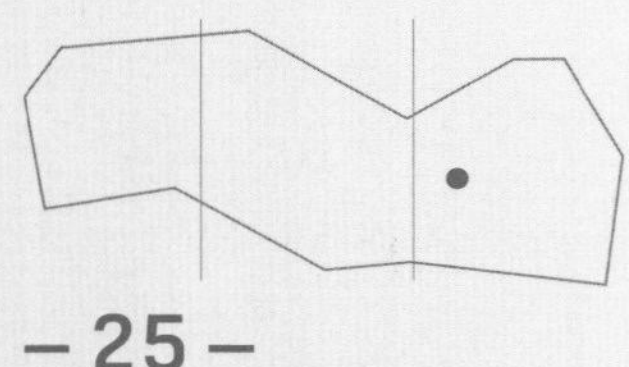

– 25 –

Galileïstraat 5 \ Rotterdam

MERWE-VIERHAVENS

Start-ups

According to researchers from the Centraal Planbureau, logistics and basic chemistry in the Netherlands are not as innovative as it they should be. These sectors are quite introvert, and have too little contact with other sectors. Crossovers are important to maintain the vitality of a sector, for example by linking major companies to the ideas of creative (starting) entrepreneurs. With these entrepreneurs, innovation can be brought further into the port. The importance of these start-ups is now widely recognized. In and around the port, incubators – breeding grounds – are popping up in different places, where young companies can be helped out with a space, knowledge, experience and networks.

One important incubator is not located in the port, but rather in Delft. That's not illogical, seeing as many civil or maritime engineers from Delft University of Technology have applied their practical knowledge in the Port of Rotterdam. Today, the ties with Delft are still very strong. The newest tie is the 'port innovation lab' at YES! Delft. Here, budding entrepreneurs are brought together. YES! Delft supports students, professionals and scientists in developing ideas into business plans. And then helps them to do real business. In this way, they are coaching the companies of tomorrow. The incubator links the start-ups to financiers and large big business in the port, which could use some innovation. Banks, governments and the Port Authority now also see the value here, and have set up special funds for start-ups and incubators that facilitate these processes.

Incubators can also be found in the port area, for example on the RDM campus. The latest offshoot is called SuGuClub and is located on the Galileïstraat in the Merwe-Vierhavens area. The location of their clubhouse is the historical workshop of the now-closed gas plant, designed by architect Ad van der Steur. The name SuGuClub suggests that something else is now going on there, with the former use of the nearby Keileweg still in mind. But SuGu actually stands for Start Up Grown Up. The initiative focuses on the circular, creative industries and aims to link start-ups that have innovative ideas to large, experienced companies.

In the old port areas, then, a renewed impetus is returning for the development of a sustainable economy.

– IV –

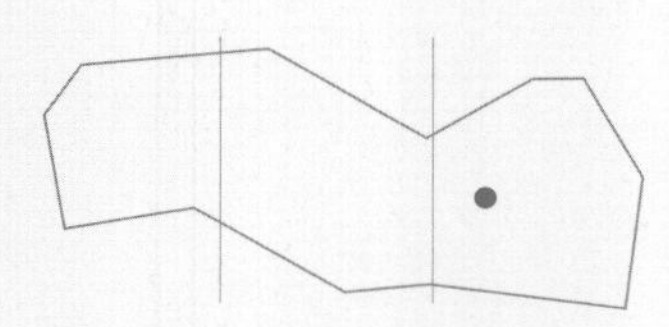

– 26 –

Keileweg 8 \ Rotterdam

MERWE-VIERHAVENS

Local Design and Manufacture

Technological innovations have led to mass production, but now the reverse is happening. The possibilities of digitization and robotics, 3D printing and sustainable energy allow you to make your own stuff. You can also produce energy in an independent and customized way. You can even offer these products and energy to the local market. It is unlikely that everyone will want to design and manufacture their own things at home; there are local companies that will do this part for you. You can also use their equipment in the workshops of these new artisans, for example the 3D printers and cutters at Makerspace at the RDM, or at Solid Shapes in the Merwe-Vierhavens area.

Economists have already announced the third industrial revolution. The technical possibilities are leading to a new way of working, organizing and sharing knowledge and information. This is also leading to new forms of financing, such as crowdfunding. This is good news for the resilience of a city or a business, as it leads to new social and economic networks.

Since 2012, local entrepreneurs have come together in the network called 'Made in 4Havens', the platform for the local manufacturing and design industry in the area Merwe-Vierhavens area. Makers, designers and artists were already present in the area, but they were quite invisible to the outside world. This network let them discover how they can complement each other and how to make connections with other companies and institutions. The initiative is expanding to neighbouring Delfshaven, and people are sent out from there on apprenticeships. There are now many renowned designers in the area, which is beginning to create something of a buzz. The local economy is boosting the area's development. The significance of this has since been discovered by developers from the municipality and from the Port Authority.

But what does this mean for the port? If the movement formed by the local manufacturing industry really takes off, this will have effects on the distribution chains. The rise of the local manufacturing industry is at odds with the one in which the port excels: scaling up, being a pivot in global supply chains and mass production. It seems inconceivable that the local manufacturing industry will replace global mass production. And whether 3D printing will lead to a bulk of raw materials for these printers coming through the port is still highly uncertain. Perhaps the printing materials will be made locally using waste from our own city. But the local design and manufacturing industry *can* reinforce mass production and make this global port smarter through the innovative and creative power that it generates. The big industrial players should be able to learn something from this..

– IV –

Urban farm 'Uit Je Eigen Stad' (From your own city).

SuGuClub.

SuGuClub.

Merwehaven.

RDM.

RDM.

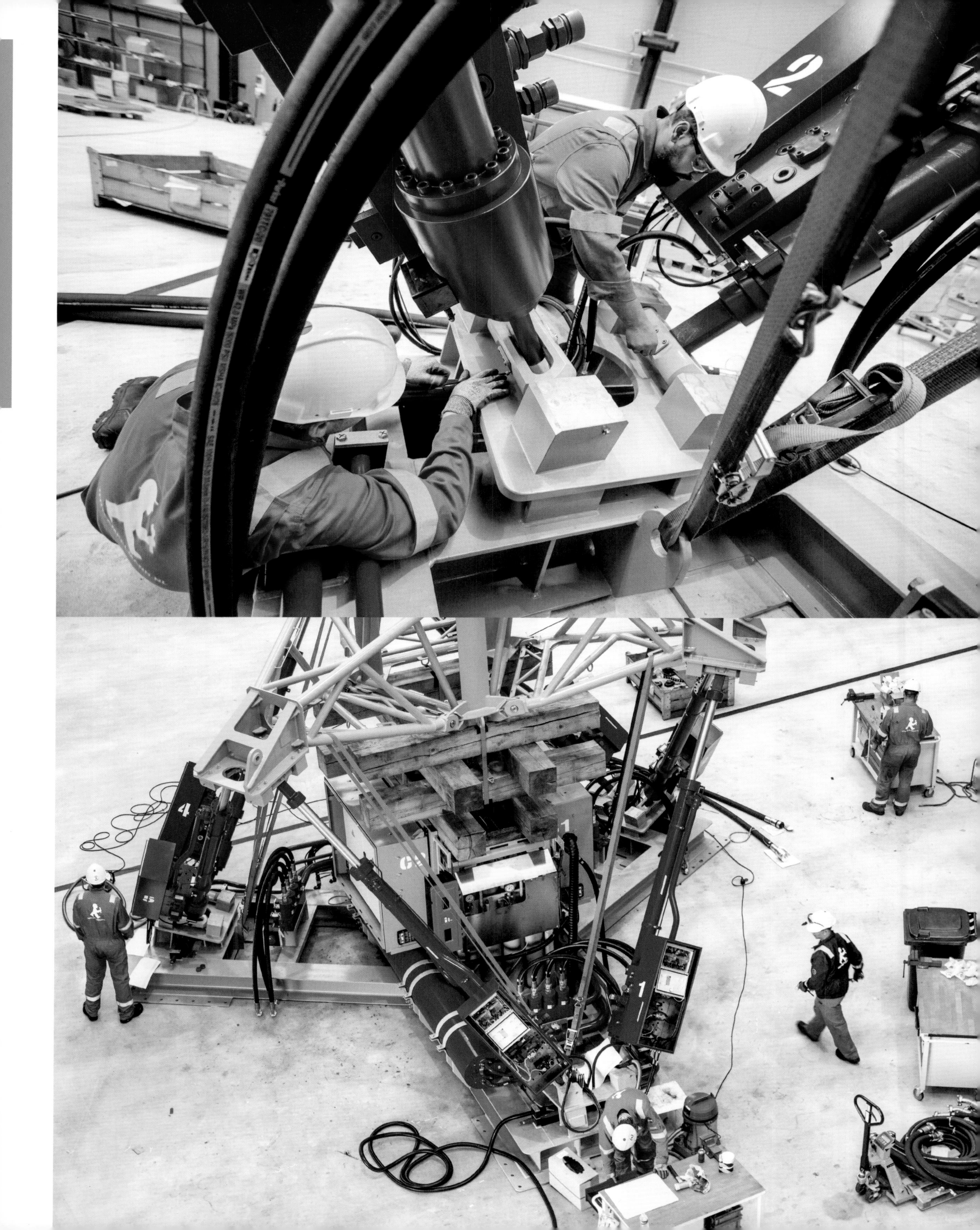
2
4
1

– 27 –

RDM-plein 1 \ Rotterdam

RDM AMPELMANN

Offshore of Course

Even though large vessels are no longer built here, the maritime industry is still very much present in the Rotterdam Delta. There are hundreds of companies here that repair, demolish, tug, salvage, dredge and so on. They are essential for the services that the Port of Rotterdam provides. But not only that. Everywhere in the world, 'we' are dispatched; it is Dutch glory.

The economic situation of the sector goes up and down, on the waves of the global economy. In the last century, brand new ships were launched here from the docks of the Rotterdamsche Droogdok Maatschappij, Piet Smit, Verolme, Wilton, Damen and the Industriële Handels Compagnie (IHC). But starting in the 1970s, the docks began to empty out. A number of docks merged into the RSV Group, which was established by the government out of fear of losing jobs. RSV closed in 1983, and some docks continued on their own. Along the water, you can still see many of the names of yesteryear. They specialize in dismantling, maintenance, repair or making specific components. The docks that went bankrupt are now popular locations for urban development on the waterfront.

The fastest-growing sector today is offshore. That also has to do with the energy transition. In the North Sea, more than 400 oil and gas platforms will need to be dismantled. The fossil resources in these spots are becoming exhausted. A large part of such a platform will have to be taken apart. The ship that will collect these materials and bring them into port, in other words the garbage truck of the North Sea, is called the *Pioneering Spirit*. This mammoth vessel, another 'largest ship in the world', is being completed in Rotterdam, and hundreds of companies are involved in this process. During its first weeks in the backwater of the Maasvlakte 2, countless day-trippers sailed by in long rows to take a look. There had never before been so many people here at the same time.

The North Sea can be a source of sustainable energy, as oil platforms make way for platforms with windmills. Offshore thus remains a thriving industry, focused on innovation. Two young engineers from Delft, for example, developed a unique concept for wave-damping platforms, which make working at sea safe and comfortable. Their product is now enjoying worldwide demand. Ampelmann has grown into a large company that has its own place in a renovated warehouse, and indeed at RDM Rotterdam, where shipbuilding used to thrive.

– IV –

– 28 –

RDM-kade 59 \ Rotterdam

RDM MACHINEHAL

Teaching Design and Manufacturing

Rotterdam is a city with many young people. Several years ago, the alarm was sounded because many of these youngsters were not given (or did not make use of) enough opportunities to finish a good education and find a suitable job. The percentage of low-skilled workers was too high. At the same time, for companies in the port, it was and still is difficult to find new people with the appropriate knowledge and skills. The port is not yet seen everywhere as a job engine. Out of sight, out of mind is certainly applicable here: the port moved westward, and became less visible in the city.

According to the Economische Verkenningen van Rotterdam 2014 (Economic Policy Survey of Rotterdam 2014), the percentage of low-skilled workers has already fallen sharply, and the proportion of highly educated ones has increased. That's good news.

In particular, technical education has begun a revival. That might have to do with the crisis, but is also the result of campaigns carried out by the government and the business sector. The courses are again filling up. Anyone with good technical training has a greater chance of finding a good job.

Educational institutions such as the Scheepvaart en Transport College, Albeda College, Zadkine College and the Hogeschool Rotterdam work hard to provide the education that matches the labour market of tomorrow. This is not easy, as implementing changes in education generally takes a lot of time. Developments such as automation, robotics, circular chemistry and new manufacturing technologies are proceeding quickly, and will lead to radical changes in the type of work that is required. By maintaining close ties to the business community, the educational field is constantly being fed by current practice.

About ten years ago, the Hogeschool Rotterdam and Albeda College presented to the Port of Rotterdam Authority their vision of a new engineering campus in the empty halls of the RDM dock. The Port Authority, as the owner, was looking for a vital repurposing of this culturally and historically valuable complex.

Today, the RDM campus is a prime example of attractive education in a magical place. Under the roof of the old machine hall, where the RDM's welding school used to be housed on an intermediate floor, students are now trained in research, design and manufacturing. They do practical research in collaboration with local businesses, for example at the adjacent Innovation Dock. Technology has become visibly more attractive at the RDM Campus. On and around the dock, everything can be tried out: a floating tree bobs on the water, there is an autarchic house, hydrogen cars drive around, and underwater drones can be controlled. This pilot factory is also a fantastic site that you can reach by boat. Technology is the future.

– IV –

2C

375.17
D
50

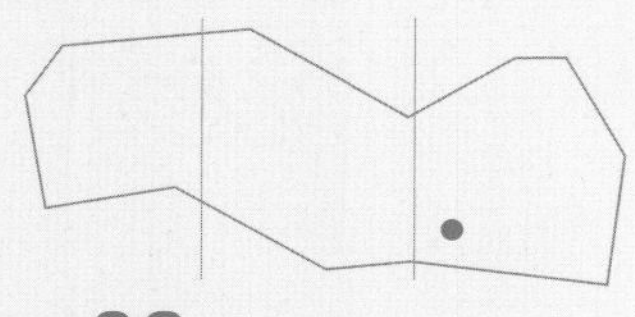

– 29 –

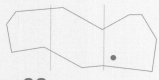

Marco Polostraat 2-14 \ Albrandswaard-Rotterdam

EEMHAVEN

The Port as a Distribution Centre

The staff is hard at work in the Neele-Vat Logistics distribution centre in Distripark Eemhaven. The Port of Rotterdam has many such distribution centres. They are scattered across the port, but are also concentrated in so-called distriparks. Goods that have already arrived in the port are stored In large hangars with the surface area of many football pitches. Distripark Eemhaven is one of the larger examples, along with the Distriparks of Maasvlakte and Botlek.

Warehousing and distribution is an important function of the port. That role has only become more prominent with the advent of the container. This made globalization possible: by producing goods in low-wage countries, cheaply transporting them by ship to the market, and from there distributing them to one or more places.

Rotterdam is primarily an import port. Yet for the Netherlands, these imports also deliver exports. Many products that enter the port are processed, either in Rotterdam or elsewhere in the country, and then exported, mostly to the European hinterland. These re-exports amount to about half of the total exports, and are thus on the same order as what the Dutch manufacturing industry produces abroad.

The processing of goods, whereby value is added to the products, often takes place in Rotterdam and the surrounding area, but that is not a given. A container can just as easily be sent on to Roosendaal, Tilburg or Venlo. Or abroad, to Liege or Duisburg. Rotterdam competes with these logistics hotspots for the favour of the distribution centres. Rotterdam has an advantage because from here, goods can easily be re-exported by sea.

Typically, it is the sender or the recipient of the goods that determines where the distribution takes place. They often leave the arrangements up to the party that they have engaged as the principal, a forwarder or logistics service provider such as Neele-Vat. 'It's a travel agency for goods,' as the forwarders describe themselves. In Rotterdam, they come in all shapes and sizes: from international concerns to local family businesses, and from general to specialized. Shippers use forwarders not only for storage and distribution, but also for the paperwork that goes hand in hand with the transport of goods: the transport documents and customs declarations.

– FdK –

CROWN

Papier
HT70
CROWN

VATlogistics

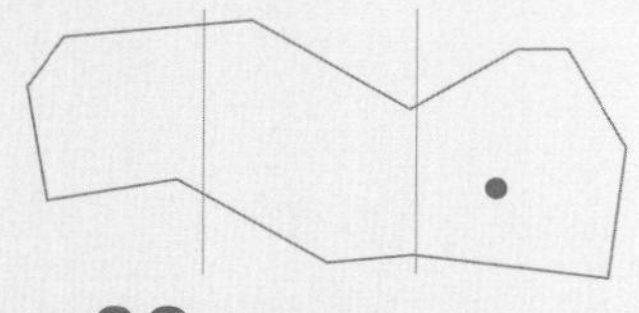

– 30 –

Waalhaven Oostzijde 1 \ Rotterdam

SLUISJESDIJK

The Borders between City and Port

The borders between the port and the city are often a bit frayed and grimy. And not only in Rotterdam. In many ports around the world, port companies are outgrowing their current sites and therefore leaving the old piers. These border locations are not usually easy to reach. You have to make your way across a causeway, a bumpy exit road or a marshalling yard. These abandoned areas are home to a variety of activities: lorry companies, suppliers, car dealers, artists, squatters, street racers, prostitutes and fishermen.

Sluisjesdijk is one such place, with a rich history. This is where the forerunners of the major refineries and technical installations began, companies such as Radio Holland, Cofely and Imtech. In 1940, the Gorzenpad workshops were built for companies that had been made homeless by the bombing. And in the 1980s, companies were put here that were forced to leave the residential areas as a result of urban renewal.

Literally on the lease boundary of the port is the old building of the Handelscompagnie, located on the Sluisjesdijk. This 1945 building had been forgotten and stood empty for years. As a result of the privatization of the Port Authority and its user, the RET, the actual owner was not even known. Parties that came forward for redevelopment wound up in a web of competing interests. That was until the Veldacademie came knocking: the local educational institution of Delft University of Technology. They had a plan that literally and figuratively linked the port and the city. The property was revived with a minimal budget, and with the support of local businesses located on the Sluisjesdijk. Students now come here to work on local issues. Entrepreneurs come to talk about future plans for the area. The long-term unemployed are sent out from here to learn the green management trade.

About ten years ago, the Sluisjesdijk was still seen as a logical sequel to the Kop van Zuid, the second Manhattan on the Maas. From the waterfront you indeed have a beautiful view of the water and of the city. But the transformation of the Stadshavens has proceeded more slowly, or at least at a different pace. Housing corporations are now focusing on renovating their current properties. The age of great master plans seems to be over. But that's okay. G.J. de Jongh's Waalhaven continues to provide space for many port companies and technical service providers. Meanwhile, the eastern side has been substantially refurbished. The challenge now is to support entrepreneurs and to connect them with the talent from surrounding neighbourhoods. This is what is happening in the Handelscompagnie's building. The question of whether this benefits the port or the city is no longer relevant, and now policymakers recognize this as well.

– IV –

DOKKANTOOR

DAMENSHIPREPAIR.COM

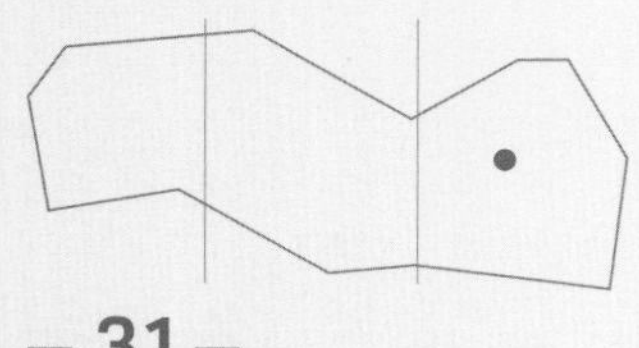

– 31 –

Stationsplein \ Rotterdam

ROTTERDAM CENTRAAL

Economy of the port city

What do the centre of Rotterdam and the port have to do with each other? At first glance, nothing. The area around the new Central Station is home to offices and urban facilities. But as the port director says with bravado: Rotterdam has grown through the port, and without the port there would be no high-speed line or Central District.

Rotterdam has since become much more than a one-sided port city, and it now appears on all the right international lists. Architecture, public space, restaurants and hotels, tourism, creative incubators. The city is attracting more and more people who want to live or work here, and these people are better educated than ever before. That doesn't happen on its own, but instead is the result of efforts by the municipality, social institutions and businesses. The development of services and creative industries in old industrial cities like Rotterdam is much more complex than it is in Hanseatic cities like Hamburg or Amsterdam, ports that have traditionally had a large service sector and a highly educated workforce.

Even though the port may be invisible, there are many port-related companies located in this city. These are the white-collar business services for the customers of the port: specialized (marine) insurance, tax and legal matters, mortgages for vessels, accounting, and commodity trading. Worldwide, Rotterdam comes in at sixth place as a maritime service centre, and it does not do much worse than metropolises like New York, Singapore or Hong Kong. London tops the list because of its logical connections to the financial world of the City. According to research from Rotterdam's Erasmus University, there are about 130 high-quality services provided here, and that number has room to grow. These companies provide around 1.5 billion euros' worth of services. Though direct employment in the port will fall due to further automation, it is expected that employment in these maritime activities will flourish. Especially now that this city is becoming more attractive as a business location.

But this does not mean that the port will become more visible in the city just yet. A decade ago, the Port Authority asked renowned designers the question: What can we do to improve the image of the port? This immediately produced a very concrete idea from the architect of the new station. The fact that the main hall of Rotterdam Central Station now boasts this port video screen is no coincidence. Some five million LEDs now illuminate the beauty of the port every day. In the city.

– IV –

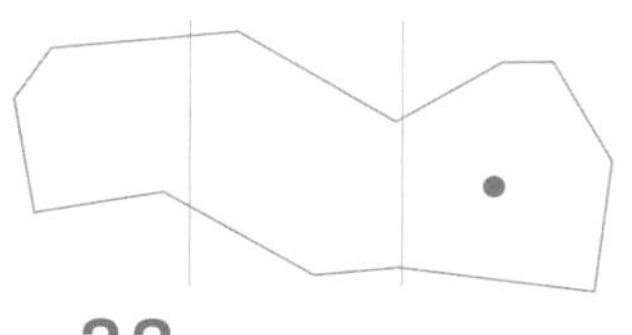

– 32 –

Veerhaven 17 \ Rotterdam

SCHEEPVAART-KWARTIER

White-collar Workers in the Scheepvaartkwartier

Lawyer Gijs Noordam looks out from his office towards the Veerhaven, the heart of Rotterdam's old Scheepvaartkwartier (Maritime Quarter). There's good reason why this district acquired such a name. This is where the ship owners and the port barons built their offices and warehouses in the nineteenth and early twentieth centuries. And a bit further down the road, they built their monumental houses. That was back in the days when the port was still located in the city, before the construction of the Waalhaven began the trend of the port moving to the west.

The transhipment of goods on the Westerkade and the Parkkade may be long gone, and the warehouses may have since been transformed into luxurious apartments, but the Scheepvaartkwartier still breathes the port. Not only because of its location on the Nieuwe Maas, or because the legacy of trade and shipping are often expressed in the architecture of the buildings built by legendary names from the past: Van Ommeren, Van Uden, Ruys en Co., the Steenkolen-Handelsvereeniging, the Koninklijke Rotterdamsche Lloyd. The district, officially called Het Nieuwe Werk, was spared during the bombing of May 1940.

The port atmosphere can also be felt because many shipping-related office functions have been preserved in this neighbourhood. The white collars have ousted the blue overalls.

Port work not only takes place on the quays, but also behind desks. Documents have to be drawn up, contracts signed, cargo insured, payments made. This does not happen only in the Scheepvaartkwartier. Right across the water, on the Wilhelmina Pier, modern office buildings have been developed, including those of the Port of Rotterdam Authority. Many shipbrokers and shipping companies can be found in the Waalhaven, including in the new Port City. A concentration of 'wet lawyers' can be found in the Scheepvaartkwartier.

Rotterdam is also where you can find the only court with a 'wet room': judges who are specialized in maritime affairs. This is also part of the port: it is not only about hunting for tonnes, but also hunting for more work in international maritime law.

– FdK –

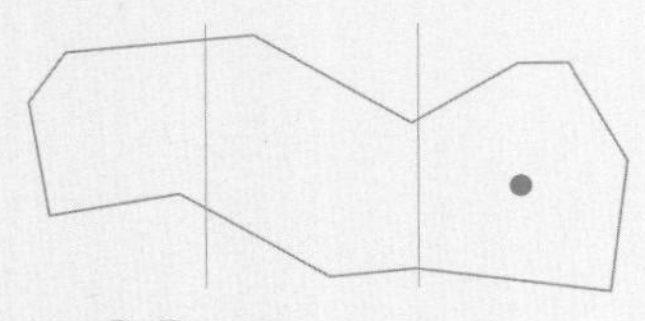

– 33 –

Wilhelminakade 177 \ Kop van Zuid Rotterdam

KOP VAN ZUID

The Maas on Fire

The building known as De Rotterdam opened in 2013, next to the Erasmus Bridge. Its architect Rem Koolhaas calls it a 'vertical city'. It houses offices, homes, hotel rooms, public facilities and a car park in one colossal building, directly on the Nieuwe Maas.

Riverbanks are popular for urban development, but new construction built directly on the quay of the Maas is not so common. The reason is safety. Ships carrying hazardous substances sail along the river. It has never happened in Rotterdam, but a ship can have an accident and lose its flammable cargo, which can then catch fire. The burning liquid can then float across the water to the quay, and then spill onto the buildings. This is called a 'pool fire', and it is why the Province of South Holland requires a minimum distance of 40 m between the riverbank and the buildings. Sometimes a distance of 25 m is allowed, but only after a specific assessment of the risk has been made.

In a building that is 150 m high, fire safety is always a high priority. But in De Rotterdam, additional measures were necessary to allow it to be located on the riverbank. It is not visible to the naked eye, but the many windows on the façade do not use the standard double glass. A special coating on the inner pane absorbs heat, and a film prevents the glass from breaking. Should the Maas ever catch fire, it would take at least 30 minutes for the flames to jump to De Rotterdam.

– PPW –

WAREHOUSING

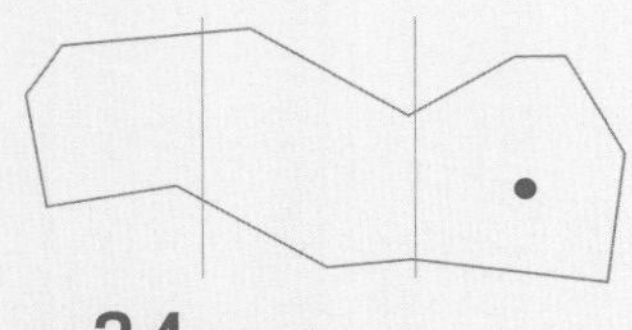

– 34 –

Maashaven Westzijde \ Rotterdam

MAASHAVEN

The 'Living Room' of Inland Navigation

This is also part of the Port of Rotterdam: the Zwarte Zwaantje playground. It is a normal playground with some typical playground equipment, but it is located on a ship, namely an old ferry from Gorinchem. Inland navigation does not really have its own place in the port, because the ships sail in and out towards the hinterland. If we had to pick a spot, that would be the Maashaven, a 'living room' that contains 160 berths for barges. Boaters and their families stay here between journeys, when they need to be ashore for whatever reason. If all is well, they only stay briefly. But in periods of economic crisis, when there is nothing to ship, the Maashaven is full.

Barges of all shapes and sizes navigate the lifeblood of the Rotterdam port, namely the rivers and canals in the interior of Europe. In particular, the Rhine is one of Rotterdam's vital arteries. The almost unlimited opportunities to reach the hinterland via the water give Rotterdam a natural advantage above any other European port. If you wanted to, you could sail along the Rhine and the Main to the Danube and arrive at Constanta in Romania, on the Black Sea.

Yet the potential of the inland navigation sector is not being exploited to the fullest. The shipping industry has long seen this mode of transport as inadequate. It is suitable for the transport of bulk (wet and dry), but not for what is now the largest flow of goods through the port: containers. After all, it's much easier to put a container on a chassis and have a lorry drive it away.

But this situation has changed over the past decade. The increasing number of traffic jams on the Dutch (and European) road system, and their associated costs, have made inland navigation a cheap and reliable alternative. The most striking example is how Heineken no longer transports its export beer to the Port of Rotterdam by road, but by water, through the Alpherium in Alphen aan den Rijn. Inland terminals like this have popped up everywhere along the waterways in the Netherlands and Europe, not least because the government has decided to promote transport via inland waterways, as it is an environmentally-friendly alternative to road transport.

For Rotterdam, this means that things will only get busier for the inland barges. The question, then, is whether the current playground for 'sailing toddlers' in the Maashaven is big enough.

– FdK –

– 35 –

Maashaven Oostzijde, pier 1329 \ Rotterdam

MAASHAVEN

Floating in the City

In the city of Rotterdam, there is really only one place where you can experience the port, and that is the Maashaven. Tough port buildings mark the south side. It has a special industrial heritage, with companies like Meneba and Quakers still making their products here. There are dilapidated palisades of grain elevators, and an incinerator (actually quite a modern one) that is ready for a second life, and which is being turned into a theme park. But above all, in the Maashaven there is an important car park and residential area for inland shipping.

The Act of Mannheim (1868) guaranteed freedom of transit and trade to the states along the Rhine, and that turned Rotterdam into a large inland port, in fact a very large one. During the industrial revolution, the steamy Ruhr area was stocked by barges travelling along Rhine. Nowhere else in the world are so many goods shipped by river barges; about half of the flow of goods in Rotterdam takes place via the waterways.

Despite that interest, the inland shipping vessels are increasingly disappearing from the cityscape. An example is the development of Kop van Zuid. Fifteen years on, the Spoorweghaven lies abandoned. And for years, the inhabitants of the Wilhelmina Pier have had a view of the Rijnhaven that is devoid of any ships. Inland shipping here was moved to the Maashaven, and what now remains is a large pool of water. Not so long ago, a site like this would have been filled in with sand, in the interest of urban development. But because of climate change and rising sea levels, now other solutions are being sought. The trend – or is it a policy desire? – is *floating communities*. Sometimes it is hard to imagine that this will ever become a pleasant sight, out there on the grey Rotterdam water. But the green engineers always come up with something: they will fill the space with floating trees!

Floating structures form a tremendously rich theme, one with a lot of space for exploration, and one where the Rotterdam delta can play a leading role, with its battle-against-the-water-history. Companies and knowledge institutes here are doing research and developing testing areas, such as Aquadock on the RDM site. These are still just experiments; the larger, more visible steps have yet to arrive in the Rijnhaven. Nonetheless, city developers are again putting the next card on the table: moving inland shipping out of the Maashaven, very much against the wishes of the skippers. Likewise, the inhabitants of Katendrecht have not asked for this change. They asked the developers for a new playground, and were given several choices. Now there is a large barge with a climbing wall on the shore.

Inland shipping remains a key success factor for a sustainable port. When the port grows, inland shipping grows along with it, which means that parking and recreational facilities are still needed. Skippers, of course, prefer to spend their free time in the city, and not in the industrial area. This is convenient for stocking up on supplies, and for keeping in touch with people ashore. In the Maashaven, this is still possible. As a result of shore-based power and new security and mooring technologies, inland shipping can be combined with urban functions. More research should be carried out in this area. It would be a shame if the inland shipping in the Maashaven were forced to make way for urban development. Anyone who wants to build a floating residential area needs to bear in mind that ships are part of the city: they are the original floating communities.

– IV –

ZWARTE
1319

ZWERVER

Meneba

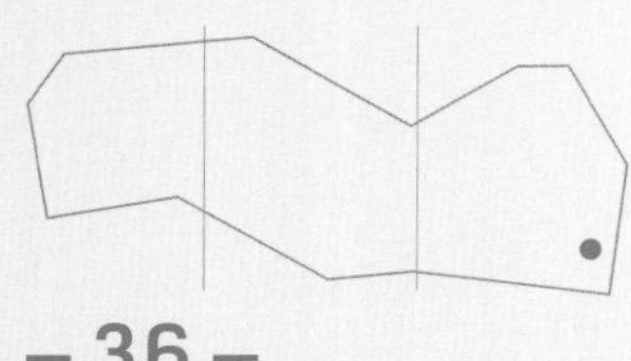

– 36 –

Verbindingsweg \ Ridderkerk

REIJERWAARD

The Fresh Food Section of Rotterdam

A container full of oranges has been placed on a lorry on the Maasvlakte; its destination is usually about 40 km inland. On the border of the municipalities of Ridderkerk and Barendrecht are the industrial estates of Verenambacht and Barendrecht-Oost. Along with Nieuw-Reijerwaard, which has yet to be built, they will form the 'Foodcenter Reijerwaard'. This is where the oranges will be unpacked and cleared, checked and processed: packaged in supermarket nets, pressed into smoothies, cut into fruit salads, and then forwarded to shops, restaurants, wholesalers and other destinations at home and abroad.

This does not take place in the port itself. Sites located along deep water are scarce. The containers are not generally opened in the port; they are moved onwards as quickly as possible. But processing the shipments creates business and employment opportunities. The Food-center Reijerwaard will employ about 10,000 people after this expansion has been completed.

In terms of policy, a strict distinction is maintained between water-related businesses and non-water-related businesses. Sites on the water are only meant for water-related businesses. Non-water-related businesses have to find a spot on the dry industrial sites inland, even if they are directly related to the port.

In 2005, the Stadsregio Rotterdam and the Province of South Holland determined that 120 hectares of dry, port-related business space was needed in addition to what was already available. It would be possible to build this extra space in the northern part of the Hoeksche Waard. The Dutch parliament blocked the plan because it would jeopardize the Hoeksche Waard's open landscape. A large section (90 hectares) was moved to Nieuw Reijer-waard, an agriculture and horticultural area near the Ridderster junction, where the landscape is considerably less vulnerable.

This move created a connection to the cluster of industrial sites located around the former fruit and vegetable auction site in Barendrecht. A revolution in logistics is taking place there, driven by rapid digi-tization and increasing consumer demand. Suppliers are now able to supply super-markets twice a day with precisely measured loads. A 24/7 economy is unfolding here.

To the north of the port is the busy Randstad, and to the south are the land-scapes of Voorne-Putten and Hoeksche Waard. Business parks like this one, where the trucks come and go day and night, are unsuitable for either the city or the open countryside. That explains the choice for Reijerwaard, hidden behind the highway noise barriers, and invisible to the general public. Maybe too invisible, because every day, many Dutch people eat vegetables or fruits that have passed not only through the port, but also through Reijerwaard.

– PPW –

VIEW FROM ABOVE

source: PDOK, 2014

Europoort with the Westland greenhouses top right.

Detail page 233.

Detail page 237.

 Botlek.

 Stadshavens.

 Detail page 239.

8+1 WORLD PORTS

Lessons for the Port of Rotterdam

Peter de Langen

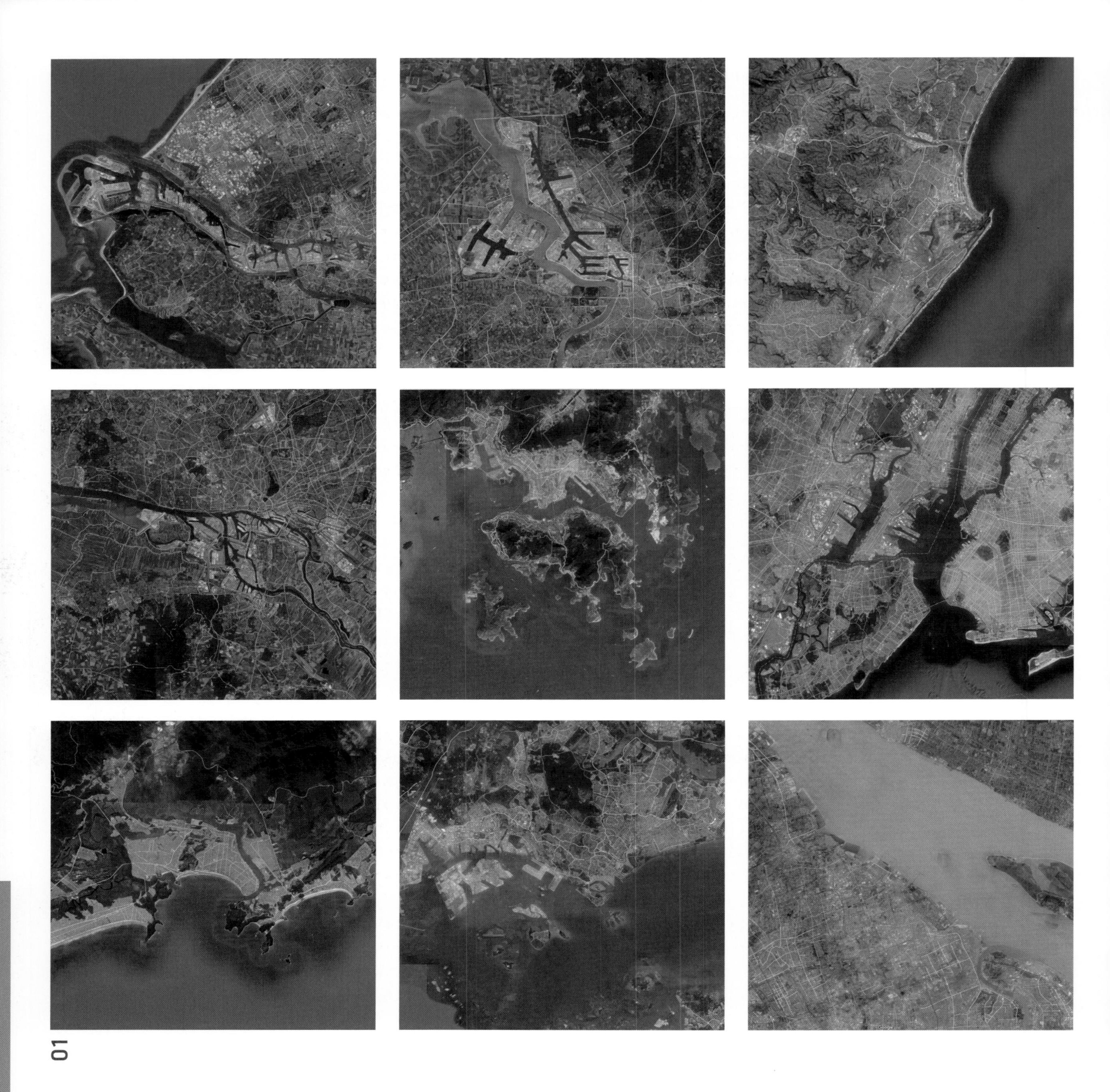

Nine port cities:
top: Rotterdam, Antwerp, Durban
middle: Hamburg, Hong Kong, New York and New Jersey
bottom: Santos (Brazil), Singapore, Shanghai

INTERNATIONAL PERSPECTIVE

'When you've seen one port . . . you've seen one port', goes the saying, to indicate how greatly ports differ from one another, in terms of size, spatial structure, development model and the type of goods that are transhipped. These differences certainly do exist, and they make the port industry fascinating. But beyond the differences, there are also similarities, and ports can learn from successful models elsewhere. That process is still in its infancy: compared to airports, which form an international industry, with international players, good benchmarks and a progressive convergence in terms of their look and feel, ports are still relatively local affairs. That also applies to the port community (in Rotterdam: Courzand, Marine Club Rotterdam), the port developer (the Port of Rotterdam Authority), and all the stakeholders that consider themselves to be involved with the port in one way or another. But the degree of internationalization is increasing. Sometimes this process comes in the form of a major shock, as happened in the 1990s, when the main container terminals in both Rotterdam and Antwerp came under Asian ownership. That shift suddenly placed the ancient sense of competition between the two ports, which is burrowed deep in their DNA, in an entirely different light.

The tour through the ports of the world in this chapter is intended to sketch an international perspective, and to show where Rotterdam resembles, or indeed diverges from, other ports. We discuss the genesis of these global ports, their port development models, the value of the port to the city and the region, and finally an important transitional challenge, and the lessons that the Port of Rotterdam can draw from it.

ANTWERP – AN ADDED-VALUE PORT

In the sixteenth century, Antwerp was the most important port and trading city in the Low Countries. The city lost this status in 1585, when the Spaniards conquered. Amsterdam promptly took over Antwerp's role as the key trade centre of this part of Europe. After the southern provinces of the Kingdom of the Netherlands rebelled in 1830, marking the beginning of Belgian independence, the remaining Netherlands sought to secure its trade interests by imposing tolls on ships sailing to and from Antwerp. British intervention on behalf of Belgium put paid to that plan, and in the century that followed, Antwerp blossomed into a dynamic port and petrochemical complex.

The amount of attention that local politics and media still pay to port issues testifies to the fact that the port heritage is deeply embedded in Antwerp's DNA. In Belgium, and particularly in Flanders, the dominant philosophy is that wealth flows into the country through its ports. Similarly to Rotterdam, Antwerp is quite a small city compared with other world port cities. The port thus has a vital role in creating value for the region's economy. Studies have shown that in Flanders, ports generate around 10 per cent of the entire economy.

It is clear, then, that port development is considered crucial. The Belgian way of ensuring that is by combining commercial ownership of the port itself with government funding of the port infrastructure. In this model, Antwerp positions itself as an 'added-value port'. Key objectives of its port development plan are the creation of jobs and added value; throughput tonnage is a less crucial indicator of success than it is for many other ports, and the task of attracting industrial and logistical operations is all the more important here,

02

02

Unloading of cargo at the Albertdok (dock) in Antwerp, c. 1930.

03

Ocean passenger ships from the *Red Star Line* at the Scheldekaaien and Rijnkaai (quays) in Antwerp, c. 1930.

04

Seamstresses restore and sew closed the cargo sacks after they are filled and weighed. Port of Antwerp, c. 1930.

05

The loading and unloading of timber carriers. Port of Antwerp, c. 1930.

06

Transhipment of goods in the Port of Antwerp. The cargo is moved from sea ship to compartments of the inland barge via slide, 1930.

03

04

05

06

07

Intersection of water ways in the Port of Antwerp, 2013.

08

Left: The church tower of the sunken polder village of Wilmarsdonk, 1967. Right: Only one of the church towers remains as a reminder of the sinking of Wilmarsdonk – it has been completely surrounded by containers, 2011.

07

08

as it is these that provide the employment and the extra value. Downstream chemistry, with products being made out of oil, is relatively more dominant in Antwerp than it is in Rotterdam or other ports. Moreover, the port lands include over 4 million m² of covered storage space, far more than any other European port.

The biggest transition challenge for Antwerp is maintaining the great economic power of port operations while combining them with improved liveability. The transport challenges are daunting in Antwerp: not only due to the presence of the port, but also arising from the region's economic dynamics and the fact that a key north-south corridor runs right along Antwerp city centre. Antwerp knows it needs to make efforts to render road haulage cleaner and more efficient, and equally urgent is a modal split: the shift away from road haulage to inland waterway and rail transport.

Lessons for Rotterdam

Whereas in Antwerp the warehouses and factories that undertake storage, added value and distribution are largely situated within the port itself, the warehouses and factories that serve Rotterdam are mainly located in the southern Dutch province of Brabant. In spatial economic terms, that model has much to recommend it: these large storage facilities are in areas with lower land prices, better employee availability and less congestion. Nevertheless, Rotterdam pays less attention than it should to distribution: the Waalhaven and Eemhaven in particular are attractive locations but there is no land to expand. That represents a missed opportunity for value creation by the Port of Rotterdam Authority and logistics firms, and missed potential in job creation.

Platform at whale factory in the Port of Durban being cleaned before the next whale is brought in, 1956.

Entrance to the Port of Durban, in the foreground the northern side of the north pier, in the background the south pier.

Rendering of the Dig-out Port project for Durban.

Arial view of the Port of Durban.

DURBAN – GATEWAY TO SOUTH AFRICA

For all cargo categories except dry bulk, Durban is South Africa's key port and is the gateway to Gauteng, the country's economic hub around Johannesburg. The port was begun as far back as the Portuguese period, but only really began developing in earnest from 1840 onwards. In more recent years, once a sandbank had been moved out of the way, Durban grew to become southern Africa's largest container port. One of the success factors here was the well-developed transit corridor between the city and Gauteng.

In Durban, just as in the rest of South Africa, the state-owned conglomerate Transnet is the dominant player in the market. Transnet incorporates the port authority (TNPA), the terminal operator (Transnet Port Terminals), the owner and operator of rail freight, and the owner and operator of the pipelines. While this is a model that has benefited South Africa a great deal, it is now increasingly inappropriate, given the changes in the port industry. Slowly but surely, Transnet is evolving from a bureaucratic government body to a commercial operation. However, its monopoly position is painfully obvious. Luckily enough for South African port users, the Ports Regular has already proven several times that it has teeth by rejecting fee rises proposed by Transnet.

Durban is the purest transit port of all the ports discussed in this chapter. The port activities do generate added value and create jobs, but not to the same extent as is seen at ports with strong logistical-industrial clusters. Consequently, it would be unrealistic for Durban to anticipate regional economic growth with port expansion alone. After all, while having a good port is a prerequisite for economic development, it is not sufficient.

09

10

12

Durban is now wrestling with the issue of how to grow: through productivity boosts or by launching a major new port development. The Dig-out Port project looks to develop new terminal capacity on the grounds of a former airfield. As is often the case, its viability depends on the growth of container volumes. The development would appear highly necessary if the very optimistic assumptions made (of approx. 8 per cent growth year on year) are accurate, but the question is whether these are likely to be realised. Should growth prove to be more modest, a more sensible decision would be to increase the productivity of existing facilities. However, the Port of Durban has limited water depth, so the shipping lines, operating bigger and bigger container ships as they are, will be quick to choose other ports where possible.

Lessons for Rotterdam

The lesson that Durban can teach Europe is that of the importance of having a good ports regulator. While that is very necessary in the African context, it is no less so in the European Union, several of whose member states (including Italy) exercise patchy oversight of the pricing and market access arrangements at ports. The Netherlands Authority for Consumers and Markets fulfils that role for the Dutch market, overseeing both the port owner and the private operators.

HAMBURG – HANSEATIC TRADITION

There has been a port at Hamburg for nearly 1,000 years, and it is the finest example of Hanseatic tradition around: a city that used development of its port to concentrate on accumulating wealth through free trade. This tradition has made the Hamburg city region one of the wealthiest parts of Europe. Hamburg's success is intimately tied to the export success of German industry. Until recently, Germany was the world's leading exporter, and even now it is surpassed only by China (whose population is, let it be borne in mind, more than 12 times the size of Germany's). Hamburg is the most specialized of all the major European ports: its business is container throughput, which accounts for around 75 per cent of its total, as compared with approx. 30 per cent at Rotterdam.

In the Hanseatic tradition, port development is a duty of the city government. With that mental map, it is no surprise that Hamburg Port Authority is so closely related to the city council. More than that, even the largest container handling company in Hamburg, HHLA, is majority-owned by the city government. The Hamburg model also extends to shipping lines and the financing of shipping; both of these are areas in which Hamburg is a world leader. Its past glories as a Hanseatic city help explain why Hamburg City Council bought shares in Hapag-Lloyd (a Hamburg based shipping line) when there was a risk of it falling into foreign hands.

The Hamburg economy rests on more pillars than just its port. Lesser-known jewels of this area include a branch of Airbus with 10,000 highly-trained, well-paid employees. Yet it is the presence of the port that explains why there is such a high concentration of Chinese trading companies in town. Hamburg is also strongly represented in high-value services connected with the port, from shipping financing to shipping line offices and consultancies.

There are two great challenges facing the Port of Hamburg. The first is that as a port – certainly when set against the local competition of Rotterdam and Antwerp, which are highly internationalized – it is still rather 'German'.

Landungsbrücken (piers) in the Port of Hamburg.

The Jungfernstieg promenade on the Alster in Hamburg c. 1900.

Tug boot in the Port of Hamburg, 1906.

View of the Zollkanaal in the Oberhafen neighbourhood of Hamburg, with the Hammerbrook quarter to the left and Großer Grasbrook to the left, 1935.

13

14

15

16

17

18

19

Speicherstadt (warehouse district) with the Elbe Philharmonic Hall in the background. Port of Hamburg, 2015.

Queen Mary 2 aside the Elbe Philharmonic Hall. Port of Hamburg, 2015.

Evening photo of the Elbe in Hamburg, 2015.

The Port Authority and the terminal operator, plus two of the key shipping lines and logistical service providers, are all German companies. This can be a strength but also a weakness now that the industry is increasingly dominated by multinationals whose headquarters are often outside Europe. The second challenge is how to retain enough support for the continuing development of the port, and here the key issue is dredging the Elbe. Given Germany's ageing population increasingly strong sentiment that economic development ought to be all about sustainability, the port's current business model is under pressure.

Hamburg has developed first-mover advantage in its use of rail freight. It would be difficult for other ports to challenge its prime position in servicing Central European markets such as southern Germany, Switzerland, Austria, Poland and the Czech Republic, where its dominance is hugely robust. There are more than 70 rail connections a week from Hamburg down to Bavaria, compared to fewer than five per week from Rotterdam. The benefits of this frequent service include economies of scale, cost-cutting and quality of service.

Lessons for Rotterdam

A challenger looking to gain a share in this market (and that is what Rotterdam is) needs patience and a deep pockets. Consequently, a more tempting option for Rotterdam is to establish a first-mover position in emerging market sectors. This could include developing into a hub for LNG as the next big fuel for shipping and inland waterways, or developing bio-based chemical industries.

HONG KONG – CHAMPION OF RE-EXPORT

The rise of Hong Kong was a direct consequence of the 1842 Treaty of Nanking, in which the Chinese Emperor ceded sovereignty of the island to Britain. This made it an attractive place to base all manner of trade-related economic activities. Hong Kong's boom years continued unabated during the tensions with mainland China's communist government, because the People's Republic depended on it to import the world's goods. Up until 2000 or so, the Port of Hong Kong remained the unassailable leader in channelling goods to and from the Chinese interior. However, the rapid development of Chinese ports in the past decade and a half, arising from the country's more open trade policy, has drastically changed the situation. Shenzhen, 100 km or so from Hong Kong, is expected to surpass the latter in terms of volumes of port trade within the next few years.

The lack of a central body for port development in Hong Kong means that Hong Kong has followed a radically different path than other ports. The terminals are private commercial enterprises, and several government bodies are involved in the various aspects of port development. It is partly thanks to this governance regime that Hong Kong is more of a collection of terminals than a truly integrated port complex. There is a Port Development Council, which advises on long-term development, but it has scant funds of its own.

As with several other major port cities, the port has in the past been a key driver of urban development here but has recently declined in regional economic importance. Like Singapore, Hong Kong makes up for that by positioning itself more as a hub for trade and maritime and logistics services. It is particularly re-export that is the strong wealth generator here now: there is a mark-up of around 20 per cent to be gained by importing goods from

25

abroad and exporting them to China proper. Thanks to this practice, Hong Kong now bathes in prosperity. It is partly for this reason that the port feels little need to blindly follow a 'growth model' through development of new port facilities.

The major challenge for Hong Kong is how to keep the port competitive and how to market the city as a business services hub. The bustling economy of this metropolis has sent land prices skyrocketing, as a result of which port-related activities – just like industry in general – have been pushed back into the hinterland. Even though Hong Kong and mainland China still have separate economic systems, the management of these port offshoots is partly done from Hong Kong. For Hong Kong, the competition posed by Singapore and Shanghai from several hundreds of kilometres away as an Asia-Pacific maritime centre is more of a worry than the competition on the doorstep with Shenzhen for container volumes.

Lessons for Rotterdam
The key lesson for Rotterdam from Hong Kong is the value that re-export can offer a port. The Netherlands is also a re-exporting country, importing goods from other continents and sending them into Europe. The re-export market actually accounts for more than 50 per cent of total Dutch exports, and that proportion has been growing for about 20 years now. A major share of Rotterdam's volumes depends on re-export sectors such as fruit and agricultural products, oil products and consumer goods for the European continent. The Netherlands' strong position as a trading country boosts the Port of Rotterdam, and vice versa.

PORT OF NEW YORK AND NEW JERSEY – A SHIFTING FOCUS

The Port of New York and New Jersey has been one of the world's greatest for centuries, undergoing a boom after Sir Henry Hudson, an English explorer in the service of the Dutch East India Company, founded a city on Manhattan named New Amsterdam. In the nineteenth century, the development of the Erie Canal was a huge shot in the arm for the city and its port, as it made the Great Lakes navigable to sea traffic. This was what let New York take supremacy from Philadelphia as the East Coast's largest city. Nobel Prize-winning economist Paul Krugman calls the excavation of the Erie Canal the decisive step that made New York the uncontested metropolis of the eastern seaboard of the United States. New York became a growth centre for port activities; this was the motor of the city's rapid development. That mechanism (self-reinforcing growth based on trade and port explains why so many of the world's great cities are, or were in the past, key ports as well.

Since 1921, port development has been in the hands of the Port Authority of New York and New Jersey (PANYNJ), a joint venture between the two states. Whereas the early organization was largely about marine port development, the focus later broadened to airport, mass transportation for the region, bridges and tunnels, and office development. It was the PANYNJ that developed the Twin Towers of the World Trade Center in the 1970s, and that commissioned the Freedom Tower after 9/11. PANYNJ's turnover from port activities is nowadays less than 7 per cent of its overall turnover.

Despite the bi-state name, PANYNJ's port operations are actually 99 per cent undertaken in New Jersey. PANYNJ is doing its utmost to hang on

26
Preparations for the departure of the first container ship to leave from the Port of New York and New Jersey, 1956.

27
Panorama of Manhattan showing skyscrapers and ships in the Port of New York and New Jersey, 1926.

28
Cranes from the Brooklyn Marine Terminal in the Port of New York and New Jersey, 2015.

29
The Port of New York and New Jersey with New York City in the background.

28

26

27

29

to New York State's tiny share by retaining port operations in Brooklyn, but despite investing hundreds of millions of dollars there, the effect is minimal. Even the port operations at Red Hook, where the Dutch began it all in 1600, are under pressure given the huge development potential (and land value) for service industries, offices and residential complexes. To some extent, the talk of a 'working waterfront' in Lower Manhattan and the dockers who are said to work there is a political fiction. Even shipping lines have moved their offices out of New York, citing the astronomical rent and wage bills. Yet New York's urban economy is not affected.

As in other major ports, New York's port developers would do wisely by anticipating the coming trends rather than clinging to the old ways of doing business. Above all, to achieve this versatility, efforts are needed to use waterways to distribute goods through the metropolitan region. It is such a densely populated conurbation that the freeways cannot cope with demand and road tolls are going up and up. Distribution costs are so high that there is a knock-on effect on prices: it is more expensive to have building materials delivered on Long Island than in New Jersey, for instance. Inland waterway transport, whether using barges or ferries, could play a role in making New York distribution more efficient. If PANYNJ succeeds in achieving this, it would deliver value to the region.

Lessons for Rotterdam

In Rotterdam, the port authority's business model is still all about tonnage and transhipment, whereas PANYNJ has diversified into office real estate (the World Trade Center) to boost its region's economy. In this way, New York has leveraged the strength of the port to develop its urban economy. Taking a similar approach in the city would help profile it as a world port city – a city boasting both a great port and a strong offering in port-related business services.

SANTOS – OPPORTUNITIES FOR PORT DEVELOPMENT

Santos is the largest port in Brazil, indeed in all South America. Port activities here date back centuries, but the first port development of substance came with the nineteenth-century coffee boom. It was then that Santos became the gateway to Brazil's commercial capital, São Paulo.

The government has a very major say in matters in the Brazilian model of port development. The same is true for several other Latin American countries. This government involvement is not delivering port efficiency: Brazil scored very poorly in the World Economic Forum rankings of port infrastructure quality. The consequences of this for economic development are considerable; the high logistic costs pose a real barrier to trade. Transaction costs are higher than they should be; administrative processes and customs procedures are far from smooth.

Santos is a real port city: everything in the urban area revolves around port operations. However, to put this in context, Santos's population amounts to less than 2.5 per cent of São Paulo's. Santos is a very clear example of a port where local revenue generation is entirely dependent on value creation in the hinterland.

The key challenge for the Port of Santos is to reform the role of the government. In broad terms, what is needed here is a turnaround from direct government involvement to a confinement of government to the regulatory

30

Sacks of coffee beans are loaded onto a ship in the Port of Santos, Brazil .

The Port of Santos, Brazil, 1929.

Ship in the Port of Santos, Brazil, 1938.

30

31

32

33

34

33 - 34
The Port of Santos, Brazil, 2014.

sphere. The current port authorities are behaving as just that, authorities, and not as port developers at the industry's service. Moreover, they lack autonomy and competence, many decisions are made at the federal level. This has brought about the frustrating situation that while Brazil has plenty of private investor willingness, the investments are not getting off the ground because they are choked by red tape and bureaucracy. Brazil would be well advised to permit more private initiative, while safeguarding the public interest. Countries on the Latin American continent that have already taken this step, such as Chile and Panama, have gone on to experience positive development and are now scoring significantly better than they used to on their port infrastructure quality indicator.

Lessons for Rotterdam

The key lesson that Rotterdam can draw from Santos is that Rotterdam's expertise is highly valuable in Brazilian port development. Port of Rotterdam Authority is active in capitalising on that expertise by developing, together with Brazilian partners, a port on a commercial basis, Porto Central. The idea is that this allows Port of Rotterdam Authority to generate really significant social value for port users. That value can be expected to yield a return in terms of shareholder value and opportunities for Dutch businesses locally. Yet, given the level of uncertainty in the Brazilian port industry, this remains a long-term ambition that calls for a long term commitment and good risk management.

SINGAPORE – GLOBAL TRADE HUB

Nearly all the world-class ports have arisen over centuries of development. Singapore is no exception, as it was already a hub for coal in the nineteenth century, when it fuelled the steamships that connected Europe via Asia to Australasia. Going back even further, Singapore was a hub for European and Chinese merchants from the fifteenth century onwards. Geopolitics caused great troughs in Singapore's economy, including the whole of the eighteenth century, when the port all but closed down. This changed in the early nineteenth century when Sir Stamford Raffles developed the island as a British port and trading post. After Singaporean independence in 1965, port development went into overdrive.

Like the city-state itself, the development trajectory of the Port of Singapore has been heavily overseen by the government. PSA (formerly the Port of Singapore Authority) has benefited greatly from the government's determination to make Singapore a commercial and logistics hub of the first order. Moreover, the people of this island, so strategically located in the Singapore Strait, have been more aware than anyone of the opportunities to be grasped from the passing trade of all shipping between East Asia and the Middle East, Europe and the eastern seaboard of North America. Due to the shape of the Strait of Malacca, the strategic advantage of this location is an aspect shared with neighbours Indonesia (which now has large-scale port development plans of its own) and Malaysia, which has swung heavily behind port development policies since 1990. Nevertheless, Singapore, having exploited the first-mover advantage, is the uncontested trading hub of South-East Asia.

It is more than just a city with a port: Singapore has developed into an international centre for the maritime industry and the commercial services around those professions. Of late, the country's government has been

35
36

35

Boats in the Port of Singapore, c. 1890.

36

Johnston's Pier in the Port of Singapore, c. 1890.

37

Evening photo of the Port of Singapore.

38

Containers in the Port of Singapore with skyscrapers in the background, 2015.

37

38

strongly promoting research and development, in the full understanding that prosperity will not continue without it. As is typical of the Singaporeans, this vision is being pursued ambitiously, with a plethora of maritime innovation initiatives at play, including a maritime innovation fund worth approximately 60 million euros.

Singapore's progress has given it a strong competitive edge, and as a consequence the terminal services and port fees are relatively pricey. This has yielded PSA a great deal of capital, which in turn has enabled its commercial expansion internationally; it is now the world's largest container terminal operator. A drop in Singapore's regional market share in the coming 20 years is all but inevitable, given the economic ambitions cherished by the surrounding nations and their substantially lower wage levels and environmental standards. The greatest transitional challenge for Singapore is therefore successful upgrading: fostering the ongoing development of high-value operations in the port, maritime and logistics sectors to avoid the economy suffering as the port's market share declines.

The Bund in Shanghai, 1933.

41
The Bund in Shanghai, 1930-1940.

Lessons for Rotterdam
Rotterdam historically has strong relations with Singapore; in the early days of their port development, the islanders viewed the Dutch as a great role model. For instance, from the earliest planning (1959) for Singaporean independence until the late 1970s, the Dutchman Albert Winsemius served as an advisor to the government. In the aftermath of the Second World War, Winsemius was directly involved in administering Marshall Aid for Rotterdam's port-industrial complex. Later, another Dutchman, Willem Scholten (a former director of Port of Rotterdam Authority), sat on the nation's pestigious Economic Development Board. However, the Netherlands would not be able to apply the Singaporean model lock, stock and barrel, as the Dutch are keen on their right to have a say and to be consulted by their bosses. Probably the most important lesson for Rotterdam is that since both cities face the same challenge of making their economy more innovation-driven, the Dutch model ought to imitate Singapore in ensuring enough vision and bold implementation to develop a world class 'innovation ecosystem'.

SHANGHAI – NETWORK TO THE HINTERLAND

Shanghai was already China's greatest port in the nineteenth century, thanks to its location at the estuary of the Yangtze, its tributary the Huangpu, and the Qiantang. Before the 1927 Chinese Civil War, Shanghai was already the biggest port in East Asia. In the first decades of communist government after 1949, the country's dwindling trade volumes meant that the port fell far behind the times. But with the turnabout in Chinese economic policy, which now focuses once again on boosting international trade, Shanghai has once again achieved staggering growth to take the accolade of being the world's biggest port.

Port development here is driven by the central and provincial government. Often, the consideration that tips the balance in favour of a development is Shanghai city region's power to affect national decision-making. Around 1990, the South China regions enjoyed considerable sway over Beijing, but more recently Shanghai has gained the leadership's ear, leading to heavy investment by central government in port and hinterland infrastructure. One of these investments was Yangshan Port (formerly Yangshan Deep-Water Port), more than 30 km out from Shanghai on two coastal islands that offer the naturally

39

40

41

deep waters that the city itself lacks. It says something about the power of the Shanghai lobby that these islands were awarded to the city council to enable the port development here, even though Ningbo (which itself is now the world's second-largest port) would in objective terms have been a more logical port to have allowed to expand its infrastructure. The key player here is the Shanghai International Port Group (SIPG), which is partly floated on the stock exchange and partly owned by the city council. SIPG has acquired a very strong position for itself throughout the Shanghai hinterland. This local dominance gives the company a solid basis for its international ambitions.

Like many a port city, there is a close interdependence between regional prosperity and port development in Shanghai. It is China's greatest business centre and owes that status partly to the power that the port has to lure banks, trading companies and ultimately the service industries around them. As in Hong Kong, the importance of the port to the ongoing development of the city itself has now taken a back seat.

Unlike Hong Kong, however, Shanghai is still a city prepared to give its all to retain its top spot as the world's greatest port. Even so, growth has been relatively mild over the past four years, as it has been throughout the Yangtze River Delta. One of the causes of this slowdown has been the very prosperity of the Shanghai conurbation itself, which has been putting pressure on industrial operations there. If Shanghai is to retain its position as the world's largest port, its main task will have to be growing as a transhipment destination that moves cargo from one container ship to another.

Lessons for Rotterdam

The key lesson here for Rotterdam is the value of having a strategy that focuses on reinforcing the network of services in the hinterland. Whereas Rotterdam faces sharp competition in the Rhine area from Antwerp, Shanghai has managed to leverage a lasting competitive advantage over rival ports by investing in hinterland ports and inland waterway operators.

42 – 43
The Port of Shanghai.

43

AUTHORS

Beukers Scholma is the graphic design agency of Haico Beukers and Marga Scholma. Together they have made many books about art, design, history, urban planning, landscape, architecture, and rezoning, often in collaboration with nai010 and SteenhuisMeurs. In the maritime domain, they have recently worked on books about the collections of the Scheepvaartmuseum, the Maritime Museum Rotterdam, and the history of Heerema Marine Contractors. In 2015, a series of stamps that they designed, featuring ship models from the Maritime Museum Rotterdam, were published by PostNL. The work of Beukers Scholma can be identified by clear and distinctive design, and is characterized by a strong involvement with the content and the editorial process.
www.beukers-scholma.nl

Frank de Kruif studied history at Leiden University. As the editor of the trade magazine *Nieuwsblad Transport*, he spent years writing about transport and logistics in general, and ports and shipping in particular. He now works as a freelance journalist and historian. In 2015, he published *Het havenschandaal. Het verbijsterende verhaal achter een miljoenenaffaire.* In the same year, he and Sjaak van der Velden published *Pensioenmiljoenen. De strijd om het pensioengeld in de havens.* Since 2014, De Kruif has been the president of the Stichting Havenman/-vrouw van het Jaar.

Peter de Langen gladly visited the port of Rotterdam as a child, often with his grandfather, and wrote an essay about it in elementary school. This explains his claim to have thirty years of experience in the port industry. He studied at the Erasmus University in Rotterdam, where he later worked, and obtained his PhD with a thesis about ports. He then worked for nearly seven years at the Port of Rotterdam Authority, and became a part-time professor at the Eindhoven University of Technology. In 2014, he founded Ports & Logistics Advisory, with which he strives to contribute to a better port industry by offering advice, analysis, and training.

Jannes Linders is a documentary photographer who focuses primarily on the cultural landscape, the city, and architecture. Recent publications that he photographed for include *In detail. Het werk van cepezed* (2013) and *Schiphol megastructuur. Ontwerp in spectaculaire eenvoud* (2013). His work has been included in the exhibitions 'Modern Times. Fotografie in de 20ste eeuw' (Rijksmuseum 2014-2015) and 'Rotterdam in the Picture. 175 jaar fotografie in Rotterdam' (Nederlands Fotomuseum, 2015). Linders' photographs also form part of various collections of documentary photography, including those of the Rijksmuseum Amsterdam, the Stadsarchief Amsterdam, and several private collections. He is currently working on a visual documentary about the redevelopment of the Sancta Maria hospital complex in Noordwijkerhout, and a project about Schiphol Airport that has been commissioned by the Nederlands Fotomuseum.

Marinke Steenhuis is an architectural historian and Rotterdam expert. As the chair of the Commissie voor Welstand en Monumenten (2008-2014), she explored every corner of the city, and for this book, she continued straight through the port. In consultation with the Quality Team of the port of Rotterdam and the publisher, she helped determine the focus of this book, and put together the concept and the team of authors. Marinke is the director of SteenhuisMeurs, a bureau for cultural-historical analysis and area development, and a member of various Quality Teams, including the team for the Afsluitdijk. She publishes internationally on the topics of twentieth-century landscape architecture and urban design, and teaches at the University of Wageningen.
www.steenhuismeurs.nl

Siebe Swart is a documentary photographer specialized in panoramic and aerial photography for landscape, urban planning, and spatial development. The built environment is also a recurring element in Swart's personal work, particularly since *Panorama Nederland*, a monumental book about how the construction of large infrastructural projects transformed the Dutch landscape (1997-2007). On the occasion of an exhibition at Huis Marseille (2011-2012), the book *Het Lage Land. Nederland en de strijd tegen het water* was published. In addition to his commissioned work, for example for the Ministry of Infrastructure and the Environment, and the Cultural Heritage Agency of the Netherlands, Swart also makes regular helicopter flights for his online archive.
www.siebeswart.nl.

Lara Voerman studied architectural history at Leiden University. Since 2004, she has worked as a researcher at SteenhuisMeurs. She also publishes in books and journals, and is a guest author at Platform VOER. Together with designer Joost Emmerik, she is working on the research project 'Landschap en representatie', funded by the Stimuleringsfonds voor de Architectuur.
www.steenhuismeurs.nl

Isabelle Vries is a corporate strategy consultant for the Port of Rotterdam Authority. She also works as a lecturer at the Hogeschool Rotterdam. Her research group carries out research into the conditions that allow for a new and sustainable economy to develop in the port city. At the Port of Rotterdam Authority, Vries is the program manager of Havenvisie 2030. She previously worked as an area developer at Stadshavens, and led the Havenplan 2020.

She has published articles about area development, waterfront development in European port cities, and the relationship between port and city; she also wrote the book *Havenzicht* (2007), which resulted from a design competition, and which formed the starting point for the 'visual quality in the port' project.

Peter Paul Witsen is an independent planologist, and since 2000 has operated as 'Westerlengte – tekst en advies voor ruimte en beleid'. Prior to that, he worked at Inro, the TNO's former institute for regional development. Witsen combines planning policy with trade journalism in the field of spatial planning, landscape, and urban design. His recent publications include *The Selfmade Land*, about the developmental course of Dutch urban planning, and *Waard of niet*, an essay commissioned by the Board of Government Advisors about environmental quality, and what the government can achieve in that regard. He writes regularly for the magazine *Blauwe Kamer*.

www.westerlengte.nl

LITERATURE

I. Blom et al., *65 jaar Havenbedrijf 1932-1997* (Rotterdam, 1997)
C. Boekraad et al., *Rotterdam 2045. Visies op de toekomst van stad, haven en regio* (Rotterdam, 1995)
City of Rotterdam and the Stichting Havenbelangen Rotterdam-Europoort, *Rotterdamsche Havenkroniek: tijdschrift gewijd aan de belangen van de haven van Rotterdam*, 38 (1938), 2
Crimson Architectural Historians, Felix Rottenberg, *WiMBY! Toekomst, verleden en heden van een New Town* (Rotterdam, 2007)
DCMR Milieudienst Rijnmond, *Integrale Rapportage Visie en Vertrouwen. Afsprakenkader borging project mainport-ontwikkeling Rotterdam* (Rotterdam, 2013)
Deltares, Ies de Vries, *Toetsing robuustheid Brielse Meer voor zoetwatervoorziening. Fase 2: definitieve toetsing*, commissioned by Rijkswaterstaat, Alterra and Evides (2014)
J. Everts, *Wandelvaart door de Rotterdamsche havens* (The Hague, 1913)
M. Halbertsma and P. van Ulzen (eds.), *Interbellum Rotterdam. Kunst en cultuur 1918-1940* (Rotterdam, 2001)
A. Karbaat et al. (ed.), *Van Pendrecht tot Ommoord. Geschiedenis en toekomst van de naoorlogse wijken in Rotterdam* (Bussum, 2005)
H. Meyer, *De stad en de haven. Stedebouw als culturele opgave. Londen, Barcelona, New York, Rotterdam* (Rotterdam, 1996)
H. Molenaar, 'Hybris in de haven van Heisenberg', inaugural address Erasmus School of Economics (Rotterdam, 1993)
F. Paalman, *Cinematic Rotterdam. The Times and Tides of a Modern City* (Rotterdam, 2011)
Port of Rotterdam Authority, *Rotterdam Europoort*, 3 (1966), 1
Port of Rotterdam Authority, *Rotterdam Europoort*, 5 (1968), 3
Port of Rotterdam Authority, *Rotterdam Europoort Delta*, 17 (1979), 1
Port Compass, *Havenvisie 2030*, established by the City Council for the City of Rotterdam on 15 December 2011
M. Provoost, 'Havenstad Rotterdam', *Archis* 10 (1997), 8-23
Scandinavian Shipping Gazette: The International Review, 12 (1938), 1
P. Van de Laar, *Stad van formaat. Geschiedenis van Rotterdam in de negentiende en twintigste eeuw* (Zwolle, 2000)
P. Verhoog (Stichting Havenbelangen), *Rotterdam. Veilig gemeerd, vlug behandeld, wereldhaven/Sicher vertäut, schnell abgefertigt, Welthaven/Havre sûr, expédition rapide, port mondial/Safely Moored, Quick Despatch, World Port* (Rotterdam, 1955)
A. Weissink, P. Lalkens, *Financieel Dagblad* (2013-2015), various articles concerning the port economy
G. Wurpel et al., *De toekomst herzien. De vervolgstappen om de duurzame haven te worden op basis van scenario's uit Grenzen aan de Groei* (Amsterdam, 2013)

Selection of other sources

Port of Rotterdam Authority Archives
DCMR Environmental Agency Rijnmond (environmental reports for Rijnmond, data 2014 and 2015)
Mapping History
Maritime Museum Rotterdam
Rotterdam City Archives

Thanks to

Nicolette Ammerlaan, Cees-Jan Asselbergs, Riek Bakker, Jan Benthem, Henk de Bruijn, Allard Castelein, Wouter van Dieren, Pieter van Essen, Adriaan Geuze, Rudolph van der Graaf, Jasna de Groot, Reny ten Ham, Abe Hoekstra, Bart Kuipers, Steven Lak, Cees van Leeuwen, Ton van der Meulen, Henk Molenaar, Geert Sassen, Victor Schoenmakers, Rob Stikkelmans, Hank van Tilborg, Steef van der Velde, Isabelle Vries, Alexander Weissink, Celi Wiersma-Barendregt

IMAGE CREDITS

Image # / Box (B)

Cover
Jannes Linders

The Port as a Landscape
1 Siebe Swart
B Jannes Linders
B Wikimedia Commons
2 © Spaarnestad Photo (Nationaal Archief/ Collection Spaarnestad/NFP)

Estimated Time of Arrival
Port Development before 1940
1 Collection 'Schieland en de Krimpenerwaard' Regional Water Authority, Rotterdam
B Maritime Museum Rotterdam, photo Erik and Petra Hesmerg
2 Nationaal Archief, Collection Maps Zuid-Holland
3 Rijksmuseum, Adolphe Braun
4 Rijksmuseum, Burkhard, A. Baedeker
5 Rijksmuseum
6 Rijksmuseum, Bert Elias Underwood, Strohmeyer & Wyman
7 Van de Laar, Van Jaarsveld, *Historische atlas van Rotterdam* (2004)
B © Spaarnestad Photo (Kees Hofker, Het Leven (LEVEN 022))
8 Van de Laar, Van Jaarsveld, *Historische atlas van Rotterdam* (2004)
9 Collection SteenhuisMeurs
10 Van de Laar, Van Jaarsveld, *Historische atlas van Rotterdam* (2004)
B Rijksmuseum
B Maritime Museum Rotterdam
11 Wikimedia Commons
12 Collection SteenhuisMeurs
13 Collection SteenhuisMeurs
14 Collection Cultural Heritage Agency of the Netherlands, Amersfoort
15 Aviodrome
B Collection SteenhuisMeurs
16 MEI Architects and Planners
B Collection SteenhuisMeurs
17 Rijksmuseum, Johann Georg Hameter
18 Rotterdam City Archives
19 Wikimedia Commons
20 Maritime Museum Rotterdam
21 Ampelmann
22 Maritime Museum Rotterdam
23 Collection SteenhuisMeurs
24 Rijksmuseum
B Collection SteenhuisMeurs
B © Spaarnestad Photo (Henk Blansjaar)
B © The Memory of the Netherlands/ Koninklijke Bibliotheek – the national library of the Netherlands

The 'Port of Rotterdam' Brand
1 Museum Rotterdam
2–4 *Rotterdamsche Havenkroniek* (1938)
5 Collection ReclameArsenaal, www.reclamearsenaal.nl
6 The Hague Municipal Archives
7, 8 Archive Stichting Madurodam, The Hague Municipal Archives
9 *Rotterdam. Veilig gemeerd, vlug behandeld, wereldhaven* (1955)
10 Aart Klein, Nederlands Fotomuseum
11 Stills *Polders voor Industrie*, Wim van der Velde (1962)
12 *Rotterdam Europoort* (1966, 1968)
13 Jannes Linders
14 *Rotterdam Europoort Delta* (1979)
15 Photo Rob 't Hart, Collection Kop van Zuid
16 Stills Port of Rotterdam, *The best of the biggest screen* (2015), Port Authority of Rotterdam

Estimated Time of Arrival
Port Development 1940 to present
1 Collection SteenhuisMeurs
B Celi Barendregt
2–4 Celi Barendregt
5 Website Historisch Rozenburg, www.historischrozenburg.nl
6, 7 Celi Barendregt
8 Het Nieuwe Instituut
B Tim Stam
9 Siebe Swart
10 Aart Klein, Nederlands Fotomuseum Rotterdam
11 Jannes Linders
12 Collection SteenhuisMeurs
13 Teychine Stakenburg, Rotterdam (1958)
14 Port Authority of Rotterdam
B Wikimedia Commons
B Rotterdam City Archives
15 Abe Hoekstra, Ton van der Meulen
16 © Spaarnestad Photo (Nationaal Archief/ Collection Spaarnestad/NFP)
17 Collection SteenhuisMeurs
B Collection SteenhuisMeurs
18 Gemeente Rotterdam
B Wikimedia Commons
B Siebe Swart
19 Collection SteenhuisMeurs
B Pinkfroot
20 Geert Sassen
21 Jannes Linders
22 Observatorium, Freek van Arkel, Hans Elbers
23 Maritime Museum Rotterdam, Maria Austria Instituut
B Port Authority of Rotterdam
B Jan Benthem

Pride, Comfort and Compassion
Adriaan Geuze on the Port of Rotterdam
Jannes Linders

Port Places
Dok 7 photo on the right Danny Cornelissen

8+1 World Ports
Lessons for the Port of Rotterdam
1 PDOK
2–8 © Antwerp Port Authority
9 NZAV photo-collection, Amsterdam
10 Wikimedia Commons
11 Africa Green Media
12 Michael Jung/shutterstock.com
13 Travelwithoutends
14 Wikimedia Commons
15, 16 Hamburg Bildarchiv
17–19 Hamburg Marketing GmbH
20–22 KITLV
23 Hollandse Hoogte
24 World maritime news
25 KUEHNE + NAGEL
26 Port Authority Archive
27 © Spaarnestad Photo (Het Leven (LEVEN 022))
28 Nieuwsblad Transport
29 Hollandse Hoogte
30 Maritime Museum Rotterdam
31, 32 Museo do Porto de Santos
33 www.ports.com
34 Diáro do Poder
35, 36 KITLV
37 Business Times
38 Ral Schamma
39–41 KITLV
42 Claudio Zaccherini/shutterstock.com
43 Anna Jurkovska/shutterstock.com

CREDITS

editor: Marinke Steenhuis
text: Marinke Steenhuis met bijdragen van Peter de Langen, Frank de Kruif, Lara Voerman, Isabelle Vries en Peter Paul Witsen
translation: Douglas Heingartner
copy editing: D'Laine Camp
photography: Jannes Linders and Siebe Swart
image editing: Marinke Steenhuis and Marloes Fransen in association with Beukers Scholma
image acquisition: Marloes Fransen
design and infographics: Beukers Scholma
lithography and printing: die Keure
paper: Magno Volume 150 g
production: Mehgan Bakhuizen, nai010 publishers
publisher: Marcel Witvoet, nai010 publishers

This publication was realized thanks to the generous support of the Port of Rotterdam Authority

nai010 publishers is an internationally orientated publisher specialized in developing, producing and distributing books on architecture, visual arts and related disciplines.
www.nai010.com.

nai010 books are available internationally at selected bookstores and from the following distribution partners:
North, South and Central America – Artbook | D.A.P., New York, USA, dap@dapinc.com
Rest of the world – Idea Books, Amsterdam, the Netherlands, idea@ideabooks.nl

For general questions, please contact nai010 publishers directly at sales@nai010.com or visit our website www.nai010.com for further information.

ISBN 978-94-6208-235-9

Printed and bound in Belgium

The Port of Rotterdam is also available as e-book
ISBN 978-94-6208-255-7

Acknowledgements

All employees of the Port of Rotterdam
Architecture Film Festival Rotterdam
Ampelmann, Willem Maijer
APM Terminals, Heleen Olland
Atelier Van Lieshout
Bob van der Vlist Photo and Party Centre
C.Ro Ports Automotive Rotterdam B.V., Jan Mahie, Sjors Bosvelt
Cargotec/Kalmar, Marco de Pater
E.ON, Ab Harpe
EBS, Monique van den Engel, Marcel de Bruin
ECT
Ertsoverslagbedrijf Europoort C.V., Marisca Gardenier, Ruth Jochems
Euromax Terminal Rotterdam
Port Authority of Rotterdam, Nicolette Ammerlaan, Henny Gilet, Jasna de Groot, Corinne van Iersel, Remco Neumann, Arie van Oord, Tie Schellekens, (patrol vessel RPA 13 and 15)
Höegh Target Autoliner
Keppel Verolme, Luciënne de Jong
Loodswezen Region Rotterdam-Rijnmond
Mapping History
Maritime Museum Rotterdam
Matrans Marine Services
MAX designers, Rotterdam
Neele-Vat, Ruud Vat
Olphaert den Otter
Quinten Buijsse, assistent to Jannes Linders
Rotterdam Academy of Architecture and Urban Design
Rotterdam Fruit Wharf, Fred Krijnen
Ryfas Helicopters
Sculpture International Rotterdam
Smickel Inn Balkon van Europa
STC, Scheepvaart en Transport College, Robbert Douma
SuGuClub, Gilbert Curtessi, Guus Balkema
Team CS
Ten Holter Noordam advocaten, Gijs Noordam
United Arab Shipping Company B.V., Dorothea Maipas, Marc van Haaren
Uit Je Eigen Stad
Veldacademie and Energy Lab Sluisjesdijk, Otto Trienekens
Waterbus/Aquabus B.V., Orlando Haynes
G.J. Wortelboer B.V., Marcel Wortelboer